SILENT
SALESMEN
BY BILL ENES

DEDICATION

To my Dad, Bill Enes, who was the ultimate collector, as he loved it all. He would have been my biggest advocate for doing this book.

Collecting coin-op was something we shared. We both started collecting machines the same year, neither of us knowing the other was also collecting. Twice we both bought the same machine, the same day, same hour, neither of us having seen the machine before; he in New Jersey and me in Kansas.

For information about future price guides or if you have some information about one of the Mystery machines or an unusual vendor contact the author directly.

Bill Enes
8520 Lewis Drive
Lenexa, Kansas 66227
(913) 441-1492

Printing: Inter - Collegiate Press
Editor: Peggy Enes
Photography Advisor: Charles Levenson

Library of Congress Number 87-91342
ISBN 0-944843-01-8

Table of Contents

Bill Enes

I thought that after fifteen years of being an active coin–op collector and dealer that writing a book on the subject would be a fairly quick and easy task. Two years later, the long list of acknowledgments is testimony to my original naivete. The twenty four mystery machines are further evidence that my knowledge is still not complete. Do any collectors out there have any info that they would like to see in the next printing? The support I received from the coin–op collector community for this book was extraordinary and deeply appreciated. It helped make this project a true labor of love.

ACKNOWLEDGMENTS

I would like to thank all of the people who contributed to *The Silent Salesmen* in so many ways.

First, I would like to thank Charles Levenson who was the entire photographic staff. He selected my equipment for me, told me how to set up my lighting, developed the film, made all the proofs, and did the final printing. He helped me out of so many jams that instead of saying "Hello" when I called from out of town the Levenson family would say, "What broke now?" I promised to thank him for not ruining more of my photographs than he did. Truthfully, he salvaged most of the worst of my disasters and made my photos look better than they actually were. Now that that book is done, maybe he will stop calling me the Task Master.

Second, I would like to thank my wife Peggy who was my entire computer staff. She was my liaison to the computer. She did this by doing most of the computer work herself. (It was quicker that way.) She typed, did page layouts, sized photographs, and proofed for many more hours than I would like to admit. I can't understand why she kept asking when she was going to get out of computer prison.

Paul Blankenmeister did a great job on the cover layout and I appreciate his guidance throughout the project.

Ken Rubin came up with the book title, *The Silent Salesmen.* Jack Benson did the art for the front cover, the art for the title page, and table of contents. Jim Northcraft did the C.O.B.'s in record time. Mark Lutz and Jim Walsh of Ottmar's Printing worked a lot of strange hours to do the camera–ready copy. Kay Walker, Sue Shultz, and Juanita Ham at Inter–Collegiate Press all went above and beyond the call of duty.

COLLECTIONS

Fred Anderson
Nancy and Peter Berganthal
Alan Bitterman
Jerry Brania
Rich Brinkos
Howie Brookstein
Joseph Costantino
Steve Cregut
Elmer Cummings
Phil Cunningham
Jay Foster
Jack Freund
Allen Gallegly
Richard, Dianne, Scott, and Jonathan Goldman
Tony Goodstone
Murry and Micki Gottlieb
Richard Graves
Jim Hance
Roger Harper
Stan Harris
John Johnson
Joe Kaminsky
Ed Kysar
Marshall Larks
Rick and Josie Lee
Steve Levine
Katie Levenson
Ken Lewis
Liberty Partners
Allen Pall
Steve Perry
Don Pom
Jim and Janet Pursell
Leonard Quinn
Joe Raney
Randy Razzoog
Don Rosinski
Ken Rubin
Dr. Smith
Dwight Strouse
Tom Tolworthy
Ernie Vigil
Greg Warren
Tracy Watson
Roger Wersksman
Mark Zarkos
Edward Zelinsky

PRICING COMMITTEE

Jerry Brania
Richard Brinkos
Jay Foster
Jack Freund
Murry Gottlieb
Richard Graves
Roger Harper
Roger Pribbenow
Jim Pursell
Ken Rubin
Curt Shawkey
Tom Tollworthy
Greg Warren
Craig Willardson

I would like to thank Heidi Levenson for absolutely nothing at all.

INTRODUCTION

The old advertisements pertaining to Silent Salesmen were not the least bit modest about the product that they were trying to sell. Silent Salesmen work 24 hours a day, they don't sleep, never come to work late or go on vacation. You don't have to feed it or pay it a salary. It doesn't drink or lie. The only thing they don't promise is that it won't run off with the merchant's only daughter.

Thomas Adams, of Adams' Gum, is commonly recognized as the man who popularized the vending machine in the United States. In 1888 he put vending machines containing his Tutti–Frutti gum on the New York city elevated train platforms. Others sought to emulate his success and "the rest is history."

The early Silent Salesmen were usually chocolate or tab gum machines. The gum industry was in its infancy and gum was only available in tabs or sticks. A German man by the name of Stollwerck was an early participant in the vending field. He assigned his patents for his beautifully crafted vending machines to the United States company, Volkmann. Stollwerck's machines vended his famous chocolate but could be used for tab gum as well. The appearance of peanut vendors just barely beat out the turn of the century.

The wedding date of gumballs and vending machines is a little vague but most probably took place between 1907 and 1911. When gumballs were introduced they were only sold over–the–counter —there were no machines designed to vend them. In spite of the confusion it is clear that pre 1907 gumball machines are of the same ilk as unicorns and the tooth fairy.

Through the years vending machines have dispensed a variety of products from holy water to toilet paper to the ever familiar gumballs. The *Silent Salesmen* deals specifically with vendors that dispensed gum, candy, nuts, and breath pellets. I further narrowed the subject matter by including only those machines whose primary purpose was vending. Many counter top and slot machines dispensed merchandise only to circumvent gambling laws.

The first United States patent for a vending machine was for a liquid drawing device issued in 1884. It is highly unlikely that searching through patents before that date is going to yield the date of the prize specimen of your collection. I have included a chart of patent dates and numbers in *Silent Salesmen because* I have personally found it to be a very useful tool in helping date machines, since patent numbers often appear on machines.

Patents contain a wealth of useful information other than the date of a particular machine including the inventor, who the patent was assigned to (usually the manufacturer), where the machine was invented and how the machine works. I was able to ascertain actual manufacturers of some machines whose heritage was in doubt by the use of patents.

The section of old advertisements is also useful in settling confusion about the names of machines, who manufactured them and who marketed them. Also included are machines that I was unable to photograph including some that I have no knowledge of existing other than in ads.

A confusing example is the Smilin' Sam from Alabam'. Characters advertised include a white man, a "colored" man, an Indian, a Chinaman, or an unpainted polished aluminum model. I have only seen a version that was painted red, but have heard of several machines that are painted black. The ad lends credibility to the argument that several colors were manufactured and that the black models may have original paint and that there are Smilin' Sam models yet to be discovered.

The section of Mystery Machines are orphans looking for their parents. I have not been able to find out much about them and I have included them in the book in hopes that my readers would help in determining their heritage. If you have information on any of the orphans contact me and I will include it in the Price Guide update.

PATENTS

This page lists both the patent numbers and the design patent numbers for each year. So now you are asking yourself, what's the difference between a patent and a design patent. Regular patents cover the area of "how it works" while design patents cover the area of "how it looks."

The following chart is very useful when you already have the patent number and you are trying to determine when a machine was patented. The number across from the year represents the first patent number issued that calendar year.

Year	Patent Number	Design Patent Number
1836	1	
1837	110	
1838	546	
1839	1,061	
1840	1,465	
1841	1,923	
1842	2,413	
1843	2,901	1
1844	3,395	15
1845	3,873	27
1846	4,348	44
1847	4,914	103
1848	5,409	163
1849	5,993	209
1850	6,981	258
1851	7,865	341
1852	8,622	431
1853	9,512	540
1854	10,358	626
1855	12,117	683
1856	14,009	753
1857	16,324	860
1858	19,010	973
1859	22,477	1,075
1860	26,642	1,183
1861	31,005	1,366
1862	34,045	1,508
1863	37,266	1,703
1864	41,047	1,879
1865	45,685	2,018
1866	51,784	2,239
1867	60,658	2,533
1868	72,959	2,858
1869	85,503	3,304
1870	98,460	3,810
1871	110,617	4,547
1872	122,304	5,452
1873	134,504	6,336
1874	146,120	7,083
1875	158,350	7,969
1876	171,641	8,884
1877	185,813	9,686
1878	198,733	10,385
1879	211,078	10,975
1880	223,211	11,567
1881	236,137	12,082
1882	251,685	12,647
1883	269,820	13,508
1884	291,016	14,528
1885	310,163	15,678
1886	333,494	16,451
1887	355,291	17,046
1888	375,720	17,995
1889	395,305	18,830
1890	418,665	19,553
1891	443,987	20,439
1892	466,315	21,275
1893	488,976	22,092
1894		
	511,744	22,994
1895	531,619	23,922
1896	552,502	25,037
1897	574,369	26,482
1898	596,467	28,113
1899	616,871	29,916
1900	640,167	32,055
1901	664,827	33,813
1902	690,385	35,547
1903	717,521	36,187
1904	748,567	36,723
1905	778,934	37,280
1906	808,618	37,766
1907	839,799	38,391
1908	875,679	38,980
1909	908,436	39,737
1910	945,010	40,424
1911	980,178	41,063
1912	1,013,095	42,073
1913	1,049,326	43,415
1914	1,083,267	45,098
1915	1,123,212	46,813
1916	1,166,419	48,358
1917	1,210,389	50,117
1948	1,251,458	51,629
1919	1,290,027	52,836
1920	1,326,899	54,359
1921	1,364,063	56,844
1922	1,401,948	60,121
1923	1,440,362	61,748
1924		
	1,478,996	63,675
1925	1,521,590	66,346
1926	1,468,040	69,170
1927	1,612,700	71,772
1928	1,654,521	74,159
1929	1,696,897	77,347
1930	1,742,181	80,254
1931	1,787,424	82,966
1932	1,839,190	85,903
1933	1,892,663	88,847
1934	1,941,449	91,258
1935	1,985,878	94,179
1936	2,026,516	98,045
1937	2,066,309	102,601
1938	2,104,004	107,738
1939	2,142,080	112,765
1940	2,185,170	118,358
1941	2,227,418	124,503
1942	2,268,540	130,989
1943	2,307,007	134,717
1944	2,338,081	136,949
1945	2,366,154	139,862
1946	2,391,856	143,386
1947	2,413,675	146,165
1948	2,433,824	148,267
1949	2,457,797	152,235
1950	2,492,944	156,686
1951	2,536,016	161,404
1952	2,580,379	165,568
1953	2,624,046	168,527
1954	2,664,562	171,241
1955	2,698,434	173,777
1956	2,728,913	176,490
1957	2,775,762	179,467
1958	2,818,567	181,829
1959	2,866,973	184,204
1960	2,919,443	186,973
1961	2,966,681	189,516
1962	3,015,103	192,004
1963	3,070,801	194,304
1964	3,116,487	197,269
1965	3,163,865	199,955
1966	3,226,729	203,379
1967	3,295,143	206,567
1968	3,360,800	209,732
1969	3,419,907	213,084
1970	3,487,470	216,419

CASH TRAY (left) **CASH TRIO** (right) These aluminum machines were originally marketed by the Reliable Nut Company. In 1947 the Adams–Fairfax Corporation, a subsidiary of Reliable Nut, took over the marketing of the company's machines. In 1948 these machines were advertised only by the Abbey Manufacturing Company. I don't know if this was just a change of name or a buy–out. The only differences are the globes and the embossing on the lids. All the companies were located in Los Angeles. These machines were available on a stand or as table models and came in penny and nickel models. The triple had a rotating, lazy–suzan base. **Abbey Mfg. Co.** 1948

ADAMS' TUTTI–FRUTTI This early 4–column vendor advertised Adams' Gum. Adams is credited with popularizing vending machines in America by putting his machines on the subway platforms of N.Y.C.

E–Z The Columbus Model D was altered to vend one gumball for a nickel for the Ad–Lee Novelty Co. and became the E–Z. It dispensed drilled gumballs with rolled up pieces of paper inside that told the purchaser if he had won an award or not. Some of the machines had marquees mounted on the lid showing the winning numbers. The machine had the large mouth front and the large side coin door. "Pat'd Sept. 15, 1908" is embossed on the front lip of the tray. The E–Z is marked "Ad–Lee" on the bottom and was available with a round or an oval globe. **Ad–Lee Novelty Co.**

AD–LEE The base of this machine is shorter than the E–Z. The mouth in front of the machine has straight sides as does the side coin door. The front lip of the tray reads, "Pat'd Sept. 15, 1908." The witness window on the side was used to discourage slugging. The globe is made of sheet metal. **Ad–Lee Novelty Co.**

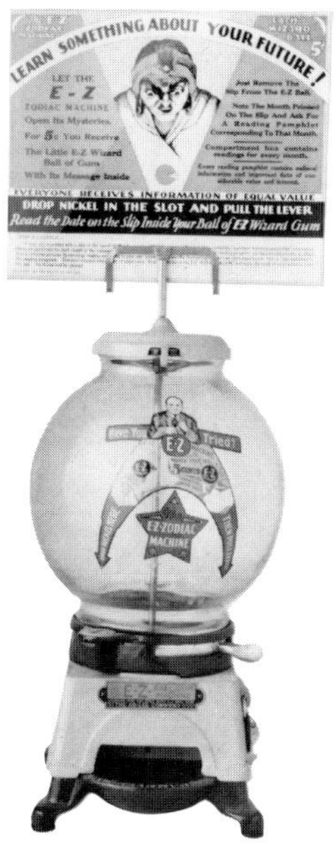

E–Z The newer E–Z is the one most of us are familiar with. The body was shortened about one inch. The large mouth opening in the front was partially covered and the side coin door was made smaller. A brass plate saying, "E–Z" was added to the front of the machine. "Pat'd Sept. 15, 1908" is embossed on the front lip of the tray. A Columbus globe without the embossed star was used for this machine. A marquee showing the winning numbers was mounted on the lid. This machine had the large E–Z decal on it. The machine was available in olive green or olive green and yellow. The machines dispensed the drilled gumballs with the slips of paper in them. The E–Z Zodiac machine had fortunes written on the papers instead of the gambling numbers, making it legal everywhere. **Ad–Lee Novelty Co.**

MODEL D This is believed to be the last Ad–Lee globe machine. It dispensed 1 gumball for the first penny, 2 for the second, 1 for the third, 2 for the fourth and 3 for the fifth generating 55¢ for every 100 gumballs. The machine was also available with 1 gumball for a penny. **Ad–Lee Novelty Co.** c. late 1920's

WORLD'S BEST MACHINE Columbus made the Model F exclusively for Ad–Lee. The peanut machine was advertised as "The World's Best Machine." Both the Ad–Lee open bar and the Columbus open star gate are correct. It was discontinued around 1929. **Ad-Lee** c. 1915

EVER–READY This is a 4 compartment penny bulk vendor constructed out of sheet metal. The machine was very similar to and apparently came after the Ad–Lee Foursome. **Ad–Lee Co.** c 1937

PEANUT MACHINE I believe this to be Advance's first peanut machine. Advance was incorporated on Sept. 20, 1900 & right after that brought out their first peanut machine. This cast iron vendor was developed by the prolific inventor, B. M. Davis. This was the start of a long line of vending machines for the **Advance Machine Co.** pat. Oct. 23, 1900.

CLIMAX The short globe on this Climax is correct. The taller globe that has ribs like a pumpkin is another scarce globe. The early Climaxes had a tray built into the cast iron base and usually had a spoked handle. An extended gooseneck type coin entry or the slug rejecter that is photographed here are both correct. **Advance Machine Co.** c. 1905

CLIMAX If your machine has the earlier Clinton St. address embossed on it, it is a clue to its age. The company moved to the Ravenswood address in 1909. This machine has been mounted on legs. Note how the face plate behind the handle is more decorative on the early models. **Advance Machine Co.** pat. 1909

CLIMAX 10 The Climax 10 is the later model and was made into the 1930's. It has four legs that are part of the base. The tray is attached and made out of sheet metal. This example has the slug rejecter coin entry. **Advance Machine Co.** c. 1915

CLIMAX 10 The Climax 10 handle is plain and solid in the center. The face plate behind the handle is plainer than the earlier model, however, the body is more ornate with triangular decorations on the shoulders. This one has the extended goose neck type coin entry. **Advance Machine Co.** c. 1915

ADVANCE JUNIOR This round globed, cast iron machine uses a modified Climax slug rejecter mechanism with a patent date on it of Mar. 8, '04. Flaps are usually found in rough condition. This model has the short body, while most of Advance's peanut machines are long-bodied. **Advance Machine Co.** c. 1904

ADVANCE GUMBALL The body is made out of cast iron and the top out of steel. It has the #4 Advance mechanism, which is dated Mar. 8, 1904 and Aug. 8, 1916. This has the small football shaped globe, but it can also use the small round globe. The stop is built into the base. **Advance Machine Co.** c. 1920

ADVANCE GUMBALL The Oct. 1936 issue of *Automatic Age* shows a route man using Advances with this globe. It has a cast iron body, steel lid, and a #4 mechanism with the patent dates 3–3–04 and 8–8–18. The spout extends out of the base with a stop at the end, like the model D. **Advance Machine Co.** c. 1936

1–2–3 Gumball The profit sharing 1–2–3 machine is quite different than the other Advance machines. It is the only Advance with a mechanism that works horizontally. The front window shows the balls to be vended before each play. It has a cast iron body and a formed steel lid. **Advance Machine Co.**

MODEL D The Model D is a gumball machine made of rolled steel and it has a nickel–plated brass mechanism showing patent dates of 11–5–12 and 4–24–22. It was available with either the small round globe or the large football globe. It has the short body. **Advance Machine Co.**

ADVANCE # 11 Nicknamed, the Big Mouth Peanut, referring to the big mouth with a swivel door that holds the peanuts until pushed, this machine has the tall body made of rolled steel. The nickel–plated brass mechanism, shows patent dates of 11–5–12 and 4–24–23. **Advance Machine Co.** c. 1925

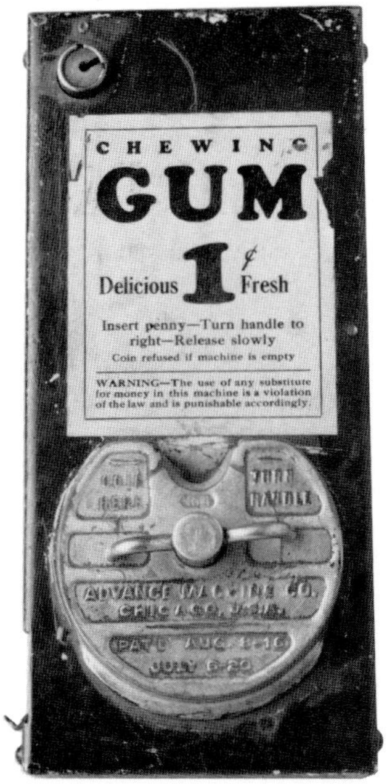

VENUS THEATER MACHINE These machines were located on the back of theater seats. Can you imagine the gum stuck on the chairs and floor? It is constructed of sheet metal with the #4 Advance mechanism which is embossed "Pat. 8–8–16 and 7–6–20." **Advance Machine Co.** c. 1920's

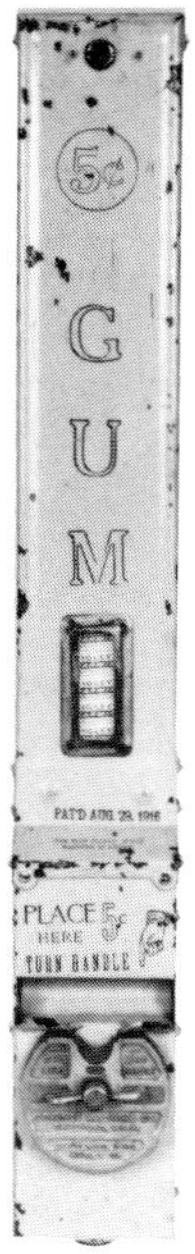

NICKEL GUM The front of this porcelain nickel gum vendor reads "Pat'd Aug. 29, 1916. It uses the #4 Advance mechanism, dated 3–8–04 and 8–8–16. The coin slot closes when the machine is empty. **Advance Machine Co.** pat. Aug. 29, 1916

AGNER & BUSHMAN'S This machine is usually found advertising Bayle's salted peanuts. The peanuts were sold in boxes. It has a great coin mechanism on the side.

Y & S I always thought that this was a unique machine, however, when I found the patent I found a few surprises. First, it is much earlier than I had originally thought, patented in 1912. Second, it was invented by William Millard who later had his own line of machines. Y & S was Alball's own brand of chewing gum. **Alball Mfg. Co.** pat Jan. 2, 1912

ALBALL Everything about this Alball is different than the other. This one has a cylindrical globe, a horizontal mechanism, is made of formed metal and is finished in a copper tiger–stripe. It was, also, made in Brooklyn instead of N.Y.C. **Alball Mfg. Co.** c. 1912

PIX SWEETMEAT GUM Albert Pick was a distributor and there seems little doubt that Automatic Sales Co. of Lansing, Michigan actually made the machine for Albert Pick. This 2–column vendor is the same as the True Blue. **Albert Pick Co.** 1906

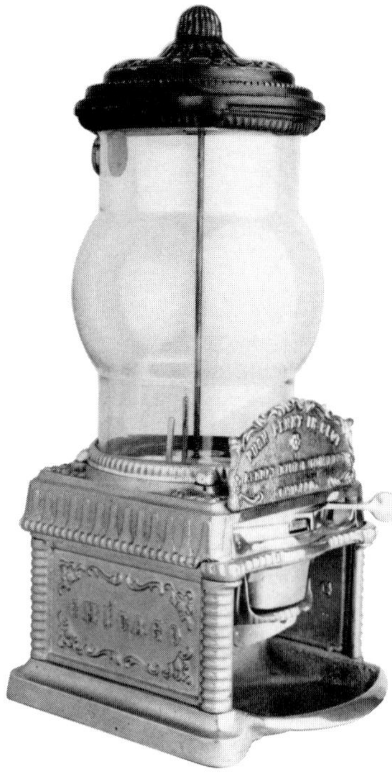

CHICAGO Coleman Hardware Co. private labeled this machine for the Albert Pick Co., a distributor. Other than having "Albert Pick Co." embossed on the front, it is the same machine as the Chicago. **Albert Pick Co.**

TRUCKY The truck drives in a circle. As the truck starts, the man on the left turns, dispensing gumballs onto a conveyor belt that unloads into the hopper. Then the truck trips the hopper, which drops the gumballs into the truck. Finishing its path, the truck drops the gumballs into a hole that dispenses them to the customer. The older version had a metal truck and operated on a penny, while the newer version has a plastic truck and gives several gumballs for a nickel. **A M Co.**

WALLACE CHOCOLATE Not only did this early, L–shaped machine dispense a chocolate bar but it also played a musical tune—all for 1¢. **American Automatic Vending Machine Co.** c. 1890's

D–LISH–US GUM This early one column vendor displayed a porcelain sign advertising D–Lish–Us Gum. There is also a variation advertising Kis–Me Gum. **American Vending Machine Co.**

KIS–ME GUM The only difference between this one column vendor and the D–Lish–Us Gum is the porcelain sign advertising Kis–Me Gum. **American Vending Machine Co.**

VARIETY SHOPS This 5–compartment penny bulk vendor is made out of sheet metal. It can sit on a counter, a stand, or be mounted on a wall. **Andrews & Co.**

ASCO ALL–PURPOSE VENDOR The All–Purpose model is a lot tougher to find than the hot nut version. It is made of brushed aluminum with the same 8–sided globe as the hot nut model. **Asco Vending Machine Exchange** 1946

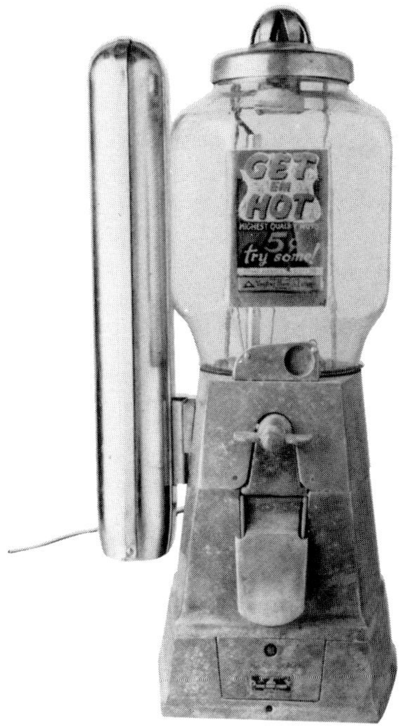

ASCO HOT NUT VENDOR This hot nut machine is made of polished aluminum. It has an 8–sided globe and a blinking light on the top. It was available with or without the side cup dispenser. **Asco Vending Machine Exchange** 1946

ATLAS ACE This is a bulk vendor that has a horizontal lever. The one–piece, aluminum body is finished in red and yellow crinkle–paint. It has a cylindrical globe with the small top opening and the small lid. **Atlas Mfg. and Sales Co.**

ATLAS DELUXE This bulk vendor has a turn handle. The two–piece, aluminum body is finished in red and yellow crinkle–paint. It has a cylindrical globe with the large top opening and large lid. **Atlas Mfg. & Sales Co.** c. 1930

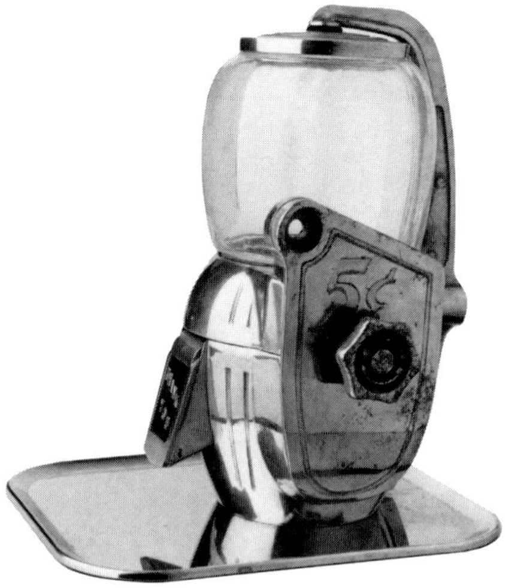

ATLAS BANTAM This machine is a 5¢ tray–type bulk vendor. The optional tray was badly needed for stability but its plating was very thin and it rusted easily. I've seen one that was embossed "Nut Bowl Jr. by Holli–Ware Mfg. Co." **Atlas Mfg & Sales Co.** 1947

ATLAS MIDGET Many collectors call this machine the Little Wonder. I don't know where that name came from. Atlas advertised the machine as the Atlas Chlorophyll Gum Vendor, the Chloromaster, and the Midget 1¢ Ball Gum Vendor. The 14–sided machine was available with or without the tray. **Atlas Mfg. and Sales Corp.** c. 1950

ATLAS MASTER In 1951 Atlas bought the Norris Mfg. Co., thus acquiring the Master name. Atlas made the Master Novelty and Master Model #2. **Atlas Mfg. Co.**

ATLAS MASTER This is a completely redesigned Master. It was a penny–nickel machine. The Atlas Master was later bought out by Buckeye Vending. **Atlas Mfg & Sales Corp.** 1954

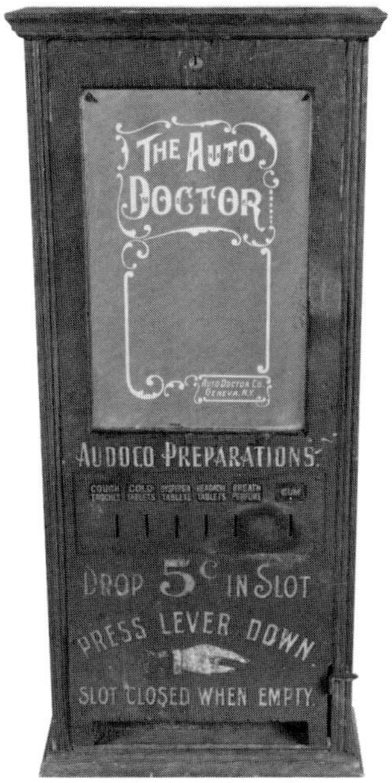

AUTO DOCTOR It's like being at your local drug store. The Auto Doctor dispensed digestive gum, breath perfume, headache tablets, dyspepsia tablets, cold tablets, and cough troches. **Auto Doctor Co.** 1892

SILVER KING DELUXE Automat Games introduced the cast iron Silver King in 1936 and offered a porcelain model for $1 extra in 1938. The flat chute cover is embossed "Silver King." Automat Games became Automatic Games in '39 and became the Silver King Corp. in '45. **Automat Games** 1938

PRIZE KING The gimmick in the early 40's was free premiums and the Prize King vended prize ball gum. The machine is the same as the Silver King except for the vending wheel. In 1939 Automat Games became Automatic Games, makers of the Factory King and Silver Charm. **Automatic Games** 1940

KING JR. Automat Games became Automatic Games in 1939. This miniature counter top bulk vendor was also known as the Professional Vendor. The doily tray was actually an ashtray and was the exception, not the norm. **Automatic Games** 1940

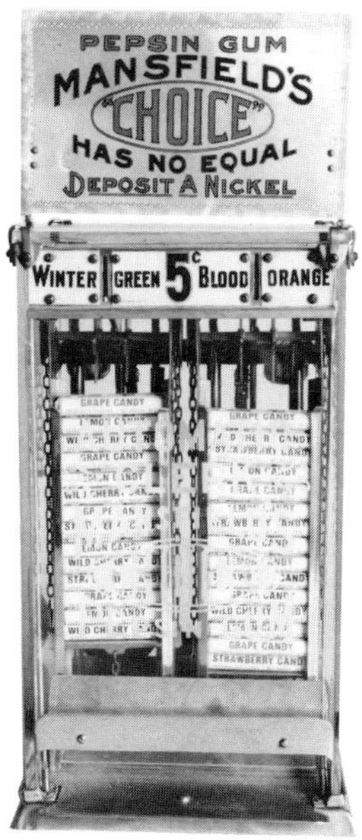

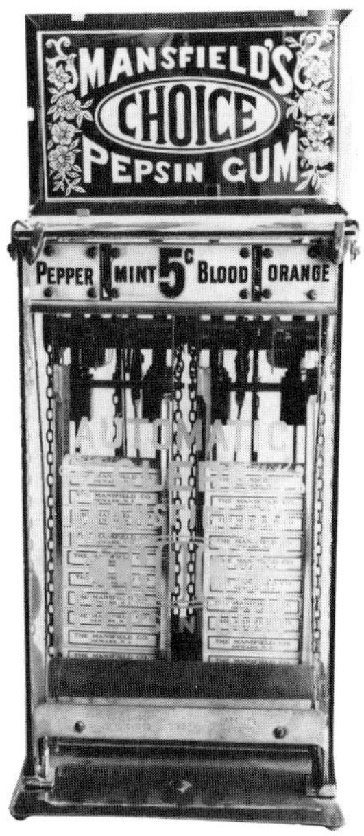

MANSFIELD AUTOMATIC CLERK For a nickel the Automatic Clerk would dispense a package of Mansfield's gum. One flavor sounds disgusting—Blood Orange! Upon insertion of the nickel, a bell rings and the gum is flipped off the top of the column. Both the beveled mirror and the aluminum marquee are correct but both have been reproduced. **Automatic Clerk Co.** pat. June 10, 1902

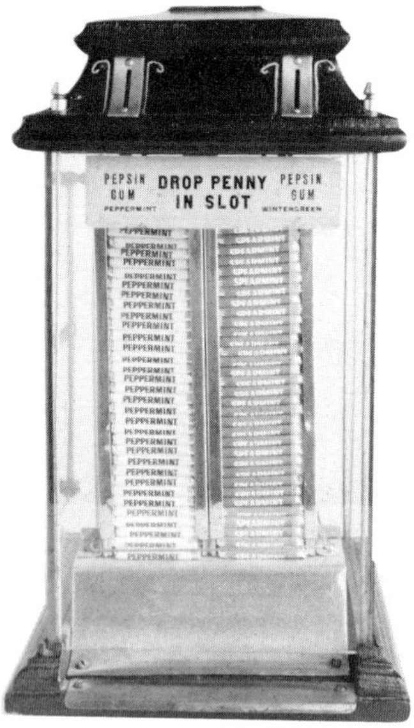

PEPSIN GUM This machine is unusual in that it has a wooden top and base. It is a 2–column tab gum vendor that dispenses the gum via a clockwork mechanism. The same company made the True Blue gum machines. **Automatic Sales Co.** pat. Nov. 24, 1903

TRUE BLUE GUM CO. I believe this to be the earlier version of a True Blue Gum Machine. The instructions for this 2– column vendor read, "Penny in Slot Push the Rod." Even with a cast iron base, it would be so easy to push the machine off the counter. It is amazing any survived at all. **Automatic Sales Co.** c. 1905

AUTOSALES CORP. The Autosales Corp. must have come after the Autosales Gum and Chocolate Co. What is interesting about this model is that the three front advertising panels are interchangeable, depending on what is in the machine. **Autosales Corp.**

AUTOSALES CORP. The Mill's Automatic Merchandising machine looks very similar to this machine, but with different advertising. It is a 4–column porcelain vendor. **Autosales Corp.**

HIT THE TARGET After the penny goes through the mechanism at the top of the machine it falls down a penny drop pin field. There was no way of maneuvering the penny; there was no skill involved. You always got a gumball, even if you got your penny back. **B & O**

CLIMAX This machine is the same as an Advance Climax except for the face plate and handle, which look like an early type "D" mechanism. The patent dates on the machine (2-11-02, 3-8-04, 7-2-07, 8-20-07, 8-27-07, and 1-19-09) are Advance dates leading me to believe they made the machine for Bayle. **George E. Bayle Co.**

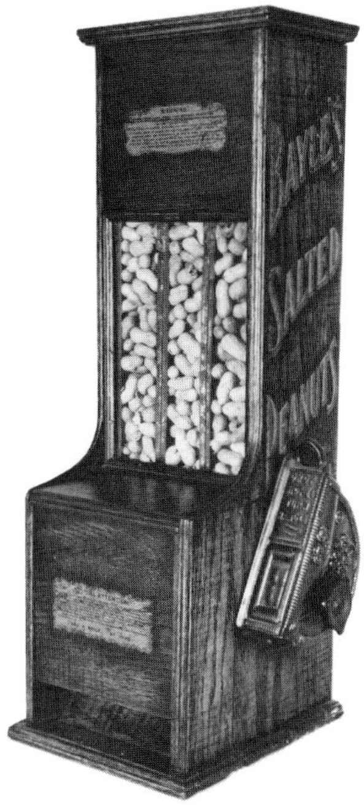

BAYLE'S SALTED PEANUTS Even though this picture shows peanuts in the shell, this machine actually dispensed 5¢ boxes of peanuts. The machine has a great embossed coin mechanism on the side. This same machine has also been seen advertising Agner & Bushman's peanuts. **George E. Bayle Co.** c. 1910

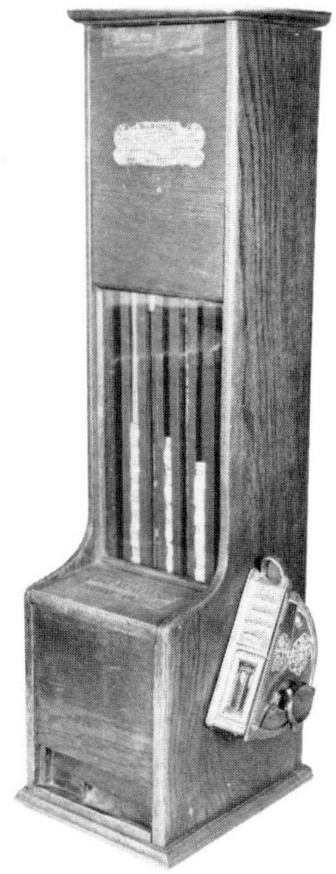

BAYLE'S Like the machine on the left this machine has also been called a Bayle's by collectors because it has the same wonderful embossed coin mechanism on the side. It dispensed 5¢ boxes of merchandise. This machine is taller and thinner than the peanut machine. **George E. Bayle Co.** c. 1910

BEAVER The Beaver is a cheaply made bulk merchandiser. The top and mid-section are metal while the base and globe are made out of plastic. **Beaver**

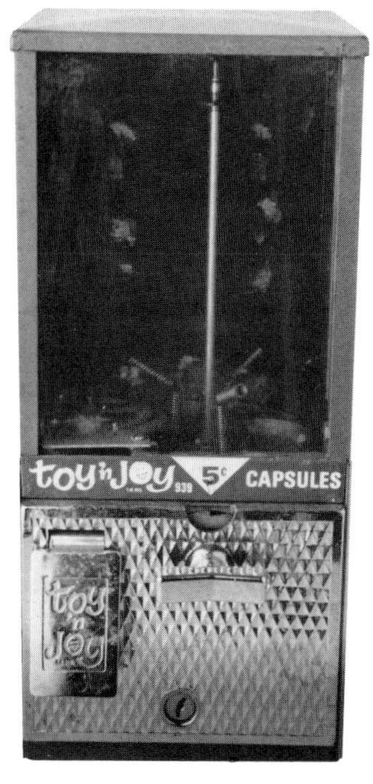

TOY 'N JOY This sheet metal machine was available in 1¢, 5¢, 10¢ and 25¢ versions. Usually found four to a stand. **L. M. Becker Co.** c. 1950's

BELVEND CHLOROPHYLL GUM Belvend was founded in 1949. This is a 3–column gum vendor made of sheet metal. **Belvend Mfg. Co., Inc.** 1950

LUCKY BOY Copper Clad, Vendor's Copy, and Bob White are other names given to this machine by collectors. It was manufactured by the Bloyd Mfg. Co and distributed by the Kentucky Gum Co. The machine is aluminum and has a tapered 10–sided globe. **Bloyd Mfg. Co.** 1937

SANITARY CONFECTIONS This is another aluminum bulk vendor made by the Bloyd Mfg. Co. This is the smaller brother to the Lucky Boy. It has a shorter body, a square globe, and a square lid. **Bloyd Mfg. Co.** c.1937

TWIRL CHUTE It's obvious why many of the collectors refer to this model as the Twirl Chute. It is considered to be the earliest of the Bluebirds. Unlike all the other Bluebirds that are made out of aluminum, the Twirl Chute is made out of tin and cast iron. It was painted gold with pin striping. **Bluebird Products Co.**

BLUEBIRD BULK VENDORS Like the Bluebird Gumball machines, the Bluebird bulk vendors have interchangeable parts. All three machines have the standard aluminum body. The machine on the left shows a built-in tray below the chute flap and the other two machines use a base that is contoured like the body and cut out beneath the chute flap. The machine in the center has a small globe and the other machines have globes the same size as the Columbus #9 globe. All these versions are correct. **Bluebird Products Co.** 1915

PEERLESS PRODUCTS CO. After the customer places the penny in the machine he maneuvers the penny chute with the control rod. Then he pushes the lever with the right hand, releasing the penny to drop through. If you hit the return trough you get your penny back. At the very least, you always got a gumball. **Bluebird Products Co.** c. 1926

BOWLING ALLEY & GUM VENDOR The bowling alley delivers 2 gumballs for every penny. The player can maneuver the shots to knock down all of the bowling pins. The machines were advertised by the Honey Dew Gum Co., which had the same address and company officers as Bluebird. Note it is the standard Bluebird machine atop the game. **Bluebird Products Co.** 1926

MAGIC VENDOR On the first machine the decal reads, “free award it the word magic appears behind the dart on the glass. Full value every time.” The center wheel has 12 numbers and the word magic on it. The second machine has a wheel with 32 numbers on it. Your chances of winning on either machine is about 1 in 32. The wheel in the center would spin each time the machine was operated. The machine was a bulk vendor and available in either a penny or a nickel version. Embossed on the flap is Peerless Products Co., K.C. Mo. Peerless used to have the same address and company officers as Bluebird, thus it is fair to assume that Peerless was a selling arm of **Bluebird Products Co.** pat. Mar. 20, 1935

BLUEBIRD 1-2-3 Bluebird gumball machines have interchangeable parts. These machines all have the large rolled base, a standard body, large midsection, large globe and a large lid which was standard with the large globe. The large midsection indicates a 1-2-3 mechanism. The large globe is the same size as a Columbus #9 globe and is the most common. The vending wheels of these machines are made of pot metal and if broken are not repairable. **Bluebird Products Co.** pat. 1915

BLUEBIRD 1-2-3 Bluebird and Simpson were the most prolific private labelers in the industry, creating confusion for the collector. A few of the many decals found on Bluebird machines are: Western States Dist. Co; Appleton Confection Co.; Supreme Products; Honey Dew; Bluebird Products; Peerless; Pamco; Universal Products; Blue Ribbon Gum; Cafeteria Confection Co.; and Cedar Rapids Gum Co. **Bluebird Products Co.** 1915

UNIVERSAL PRODUCTS Bluebirds consist of interchangeable standard cast aluminum parts. This one has the small vertical base, a standard body, a small midsection, a small globe, and a small lid which is standard with the small globes. A small midsection usually meant 1 gumball for a penny, however, Bluebird did make a 1-2-3 small midsection machine. **Bluebird Products Co.** pat. April 6, 1915

GUM When I initially thought the window located above the coin entry was a gambling window. Actually, this is a 1-2-3 machine and the window tells how many gumballs the customer will receive for his next penny. The rest of the parts are the standard Bluebird parts: small top; mid-sized globe; tall 1-2-3 mid-section; standard body; and large base plate. **Bluebird Products Co.** pat. April 6, 1915

GUMBALL What is different about this Bluebird is that it has a trap door at the end of the chute. The rest of the machine consists of interchangeable standard cast aluminum parts: small top; small globe; small midsection; standard base; and large base plate. **Bluebird Products Co.** pat. April 6, 1915

AUTOMATIC CANDY JAR The name says it all, this machine looks just like a candy jar sitting on a Bluebird base. The machine was advertised by the Honey Dew Gum Co. which was located at the same address as Bluebird. The officers of each company were identical as well. **Bluebird Products Co.** 1928

GUM This machine is the same as other Bluebirds from the body up, however, this machine has an 8–sided tapered body. Bluebird had three standard–sized globes for their machines and this machine has the middle size. It also has the tall midsection, indicating a 1–2–3 machine. **Bluebird Products Co.** c. 1918

BRICE WILLIAMS The cabinet of this machine closely resembles a Masters, however, this one has a cast iron base and a porcelain tray. The coin mechanism is unusual in that a nickel gets five turns to the left while a penny gets one turn to the right. **Brice Williams** pat. Nov. 20, 1921

BUFFALO GUM This is a large domed machine with a clockwork mechanism. After the penny is inserted, the four columns rotate one column to the left. I have seen this machine with different shaped signs on the finial: a buffalo; a rectangle; and a star. **Buffalo Gum Co.** Pat. June 4, 1907

CALVERT VENDOR The adjustable reward mechanism allowed for 2¢ to 5¢ trade payouts. A nickel denomination was also available to vend drilled, numbered gumballs. The mechanism, located on top to eliminate corrosion, shows the last three coins. In 1937 Townsend Mfg. Co. became Calvert's successor. **Calvert Mfg. Co.** c. 1935

BANTAM Nicknamed the Beehive, this machine was available with either a bracket or a chrome tray. The optional tray was advertised as an ashtray or to catch waste (sounds disgusting). It came as either a bulk merchandiser or a gumball machine. **Will P. Canaan** c. 1940

CARLTON ROCKET The machine is entirely made of plastic except for the vending wheel and the chute cover, both of which are made out of metal. The Rocket was a promotional machine. The operator would get one free after purchasing a said amount of gum. **Carlton Candy Co.** c. 1950's

CASE PEPSIN GUM The chromed base supports a beautiful bulbous glass globe that houses two columns of tab gum. **Case Chicle Co.** c 1904.

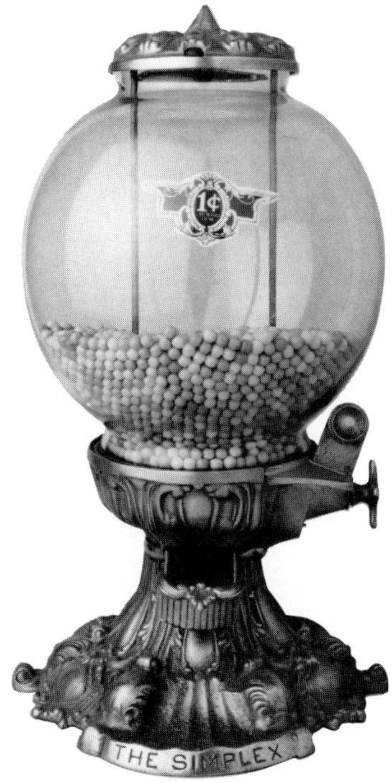

SIMPLEX This machine looks exactly like the fancy Leebold. The only differences I could find are the front is indented with "the Simplex" and the back is embossed with "Cast Mfg. Co., Indianapolis, Ind." **Cast Mfg. Co.** c. 1920

CEBCO HOT NUT A cast aluminum machine with two globes so you could vend two types of hot nuts. The nuts were heated by a light bulb. This machine also has a side cup dispenser. **Cebco Products Co.** c. 1930's

NATIONAL SELF-SERVICE MACHINE This 6-column vendor dispenses nickel packages of mints or gum. I have also seen this machine with the Gem and Bill Fryer Assoc. names on them. **Chicago Gum and Candy Co.** Pat. Mar. 26, 1925

CHICLE-CHOCLE BALL GUM GIFT BOX This machine is not exactly for tight spaces as it measures 19"×19"×14". Lettered balls were used to stimulate the sales of Chicle–Chocle Gum by the use of free gifts. The "A" ball received a candy bar and each successively lettered ball gave a neater prize, up to the "G" ball, which won a free air rifle. **Chicle–Chocle Products**

LANGLEY'S "This machine is licensed for the exclusive use of Chiclemint Chewing Gum. The use of any other gum therein is prohibited by law" is embossed on the globe of this machine. The machine is constructed of pot metal. **Chiclemint Corp.** 1915

CHIC–MINT The Chic–Mint is exactly the same as the Perfection. This early machine is made of sheet metal and cast iron. **Chic–Mint Gum Co.** c. 1915

CHIC-O-BERRY A mechanical bellows makes Mickey Mouse jump up and down and squeak. I have heard of a Donald Duck and Pluto also, but I have not seen them. They were not authorized by Walt Disney. While most of the cases were a red crinkle paint it was also available in green. c. 1930's

CLAWSON CANDY MACHINE Penny stick candy or rolled gum was vended by this machine. It is constructed out of cast iron and has a tiger–striped copper finish. Clawson used a shorter version of the machine to dispense pencils for a nickel. **Clawson Machine Co.** pat. Oct. 24, 1917

FOOTBALL–BASKETBALL–BASEBALL There is some confusion about the manufacturer of this machine that came in several sports themes. Some of the machines are marked "CVI" which stands for Coast Vending Inc., but some collectors call this a Windsor. The object is to flip the gumball into one of the holes. It has a sheet metal base and a plastic globe. **Coast Vending Inc.** c 1950's.

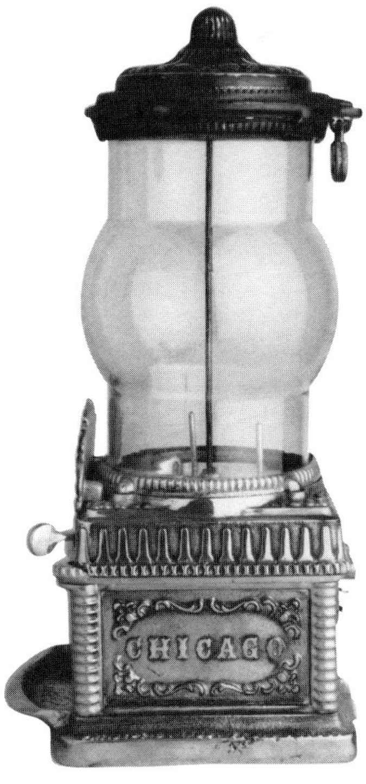

CHICAGO This is another great machine patented by B. M. Davis and assigned to the Coleman Hardware Co. Sun, Star, and Morris are other names that I have seen embossed on the side of this machine. One version has "Albert Pick" embossed on the front. While doing research I found a patent for a round base variation. Has anyone seen this version? **Coleman Hardware Co.** pat. Sept. 8, 1908

SIMPSON I believe this machine, embossed "The Simpson," is the forerunner to all Columbus machines. R. D. Simpson invented the machine with the Sept. 15, '08 patent date on it. This machine is the same shape as the patent machine, however, a spacer ring has been added between the base and the midsection. The lid, with cam lock, rotates onto a threaded rod and the coin entry are the same as shown on the patent. An open gate flap should be on this machine. c. 1907

MODEL A This is a very early Model A. It has the open face gate that the health department declared unsanitary. The gates with the star will not fit this machine. The base is a little larger than the newer versions. The older Model A's were painted black with pinstriping—red wasn't available until around 1910. There is no patent date on the narrow–lipped, built–in tray. The Sept. 15, '08 embossing on the lip came about later. **Columbus Vending Co.** 1908

MODEL A This is the classic vending machine that everyone has in mind when buying their first machine. It is a cast iron machine with a round globe. The flap has either a stencil–cut or solid star on it. Barrel locks did not come into use until 1933; until then regular locks were used. **Columbus Vending Co.**

MODEL A This is the Model A with a tray that is seen the most often. The tray has the wide lip that is embossed with the patent date, "Sept. 15, 08." This machine was discontinued around 1920. **Columbus Vending Co.** c. 1915

MODEL A This is an early variation of the columbus Model A with a tray. The patent date is embossed on the side of the base instead of on the lip of the tray, which is narrower than a normal Model A, and the base pedestal is fatter. The slug detector on this machine is of a 30's vintage. **Columbus Vending Co.** c. 1910

MODEL A This is a Model A sporting a slug detector manufactured by someone other than Columbus. The detector is circa 1920's. **Columbus Vending Co.**

HOT P–NUT The top and heating element of this cast iron machine are made of aluminum. The lid reads, "Hot P–Nut Machine Co., Columbus, Ohio". The paper decal states, "Columbus Model A Made Exclusively For Hot P–Nut Machine Co. **Columbus Vending Co.** c. 1915

MODEL B It is built exactly like the Model A, except it has a larger vender pocket and chute cover because it operates on a nickel. It's nickname is the Long Gate. **Columbus Vending Co.**

MODEL B Both of these machines are also Model B's, however, they have a tray molded into the base. Embossed on the lip of the tray is "pat'd Sept. 15, 08." What is unique about the machine on the right is that it is finished in green porcelain. **Columbus Vending Machine Co.** 1910

OHIO VENDOR This is also known as the Columbus Model C and is the forerunner to the Model D. This square–based machine had runners on the bottom instead of legs. It has a large mouth front and a large side coin door. Oval and round globes are both correct. **Columbus Vending Co.** 1910

MODEL D In 1911 the Common Sense Gum Co. started making chewing gum in the shape of a ball and they asked Columbus to make them a machine to dispense the gumballs. The model C was altered to vend one gumball for a penny and became known as the model D. Legs had been added to the body and it had the large mouth front with the large side coin door. The early model D's did not have the date embossed on the front lip of the tray. After several years the Common Sense Gum Co. ran into financial difficulty, and discontinued the use of the model D. The machine was then offered to other jobbers. Both the large Columbus globe without the star and the elliptical (football shaped) globe are correct. **Columbus Vending Co.** 1911

MODEL E The Model E is a cast iron breath pellet vendor. It has mistakenly been called a salesman's sample. (Wouldn't it be neat if it really was.) All of these model E's I have seen have a ribbed globe. **Columbus Vending Co.** c. 1909

MODEL E This is a different cast iron Model E breath pellet machine. This one has a tray. "Columbus Model E" is embossed on the base. This model has been found with clear globes without ribbing. **Columbus Vending Co.** c. 1909

MODEL F Both the open bar gate and the Columbus open star gate are correct for this cast iron peanut machine that was made exclusively for the Ad–Lee Novelty Co. **Columbus Vending Co.** c. 1915

MODEL JM There does not seem to be enough of these to go around to all of the Columbus collectors. The body is shaped differently than any other Columbus. The body and lid are light green porcelain. The raised coin entry is actually a slug rejecter. **Columbus Vending Co.** pat. 1934

MODEL K The Model K was the forerunner to the Model 14. It delivered one ball of gum for either a penny or a nickel. The midsection is smaller than the Model 14. **Columbus Vending Co.** 1917

MODEL L The pear –shaped sad face of the Model L does not move like a Blinkey. However, there was a Columbus Model G, with a clown face with moveable eyes and mouth. The Model G advertised Adams' Tutti–Frutti Gum. **Columbus Vending Co.** 1910

MODEL M or MG (left) **MODEL M** (center) **MODEL MB** (right) The first models with the bell bottom base. There were many later variations of the Model M. The "M" is a peanut machine and the "MG" is the gum model. The standard globe is the no. 4 octagon (shown in the center), but the #8 and #9 globe were also available upon request. The "MB" is a nickel version with a deep coin pocket. These machines were available in porcelain, painted cast iron and in chromium plate. They were later manufactured in aluminum. **Columbus Vending Co.** 1928

DART or MODEL R With each play, a gumball is dropped and held behind the witness window above the delivery spout, while a wheel with colored darts spins. The player wins if the gumball matches the color of the exposed dart. The gumball is released after each play. **Columbus Vending Co.** 1933

OK VENDOR or MODEL RP Advertised as "better than a punch–board!", this 5 cent gum vendor held 1200 gumballs numbered from 1 to 1200. The machine would take in $60 and pay out $30 in trade awards. The RP is an R without a wheel or witness window. **Columbus Vending Co.** 1936

MODEL RW The "RW" does not have the dart wheel, but the rest of the machine is the same as the Dart, including the witness window. **Columbus Vending Co.** 1933

MODEL V This model dispensed gumballs, marbles, or a metal shell called the "Sphero–Container" which usually contained aspirin. The #293 glass cylinder globe was not interchangeable with other models. **Columbus Vending Co.** 1930

COLUMBUS TRIPLE This machine was virtually unknown until recently. It is constructed out of aluminum. The lights on the top are for attention only and are not meant to heat the nuts. The three machines sit on a common base plate that has three separate coin doors underneath. **Columbus Vending Co. c. 1935**

COLUMBUS MODEL 14 (left) **MODEL 16** (center left) **MODEL 17** (center right) **MODEL 18** (right) If all of these machines look similar, it's because they are. They look like a Model A with a slanted coin entry. The Model 14 is a profit sharing gumball vendor with a 1–2–1–2–3 mechanism. It took in 55¢ for each 100 gumballs vended. The Model 16 vends two gumballs for each penny. The Model 17 vends one gumball per penny, The Model 18 vends one gumball for a penny, like the Model 17, except it had a larger vending wheel for a bigger size of gum. **Columbus Vending Co.** 1923–1924

MODEL 21 This small porcelain peanut machine was designed to be used on tables in taverns or other places where space was limited. It was extensively advertised with lamp and ashtray attachments. The #3 octagon globe was the standard globe and fit only the Model 21, Model 34, and the Mystery. The slug detector on the machine is unusual due to the flush mounting. **Columbus Vending Co.** pat. 1934

MODEL 34 This is a small porcelain or cast iron gumball or marble vendor which used a # 3 globe. It only fits a Model 34 or Model 21, but is not interchangeable with any other Columbus. A larger #4 globe can be used. **Columbus Vending Co.** 1936

MODEL 34 GAMBLING The player compared his numbered or colored ball to an award card to see if he was a winner. The aluminum body is based on the 34 casting. The base plate is deep to accommodate quarters. **Columbus Vending Co.** 1936

BI–MORE The model 32 originally came in light green porcelain or cast iron. It was a table top model for locations where more than one type of merchandise was desired. Columbus literature says that the octagon globe was the norm but it was available with the #8 round globe. The aluminum Bi–Mor on the sheet metal stand was the model 39 and came about later. **Columbus Vending Co.** 1936

TRI–MOR or MODEL 38 In order to compete with the other multi–compartment bulk vendors on the market, Columbus introduced the Model 38. It was advertised as having a modern streamlined design. The machine was made of aluminum and the base was made of sheet metal. The standard globe was the No. 4 octagon type. **Columbus Vending Co.** 1939

COLUMBUS 36 The Columbus 36 was a cast iron penny stick gum vendor. Supposedly, the number 36 does not represent a model number but the number of sticks of gum that the machine holds. It operates like a match vendor. **Columbus Vending Co.** c. 1915

MYSTERY A cam lock lid and a normal lid were available for this cast iron machine with embossed Art Deco sides. It takes the #3 octagon globe, seen with and without the star, that also fits the Model 21 and 34. Some have been seen with Minnesota Mining and Mfg. Co. on decals. **Columbus Vending Co.** c. 1930's

LAST COLUMBUS This machine is believed to be the last Columbus. They were made of aluminum and came in a peanut and a gumball version. The globe is shaped like the letter "D" (flat on the back and curved in front). Columbus went out of business in 1963! **Columbus Vending Co. c. 50's**

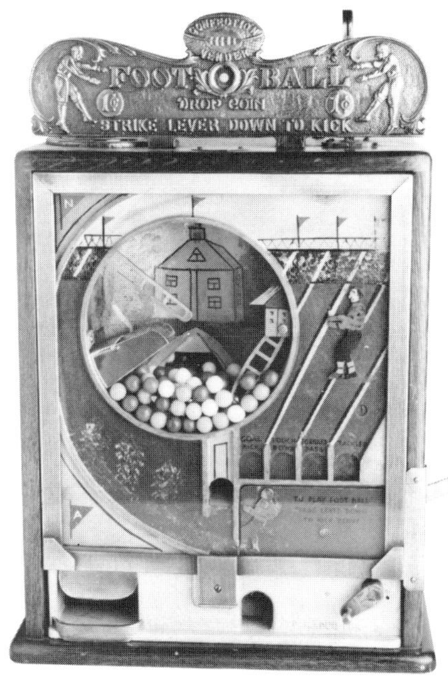

COMBINATION VENDOR The Combination Vendor has four sections. The two inner columns are for tab gum while the two outer compartments vend bulk merchandise. **Combination Vendor Co.** pat. Oct. 25, 1904

FOOTBALL Press lever to kick the penny and turn the handle to move the upper man to guide penny into the goal kick, touch down, forward pass , or tackle slot. The gumball is then raised by a scoop and goes down a chute. This machine has a wood case with a copper–flashed cast iron front. **Confection Vendor**

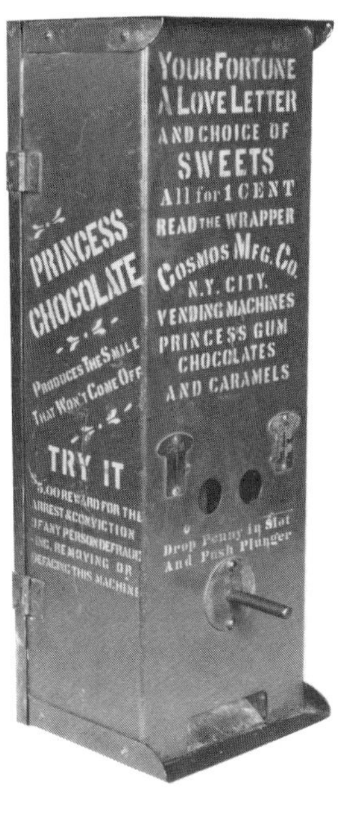

PRINCESS GUM The customer received his fortune, a love letter, and his choice of gum or chocolate all for a penny. The 2 column vendor had gum on one side and chocolate on the other. The windows told you if the column was empty or not. Note reward. **Cosmos Mfg. Co.** c. 1910

JAW TEASERS The plastic machine was made by United Metal Co. for Cramer. It was given away with large orders of gum. The machine was meant for a store as it was not sturdy enough for an operator to run on a route. **Cramer Gum Co.** c. 1950's

B–WISE When you bought their gum, you got the machine free. The gum was available in cases or in a tin that had lithographed gumballs on it, like the top of the machine. The plastic machine was made for Cramer by the United Metal Co. **Cramer Gum Co.** c. 1950's

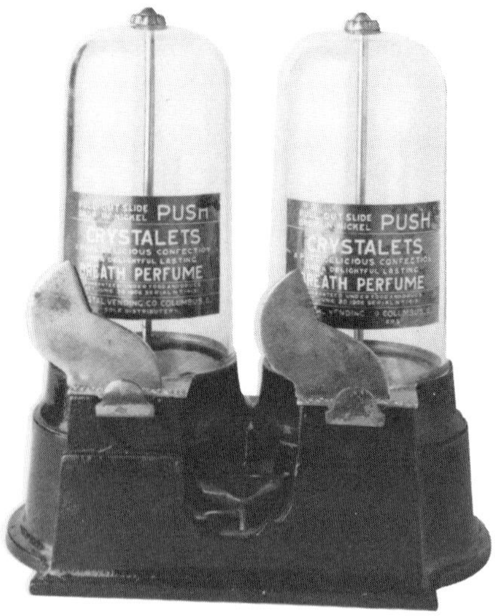

CRYSTALET MACHINE This cast iron breath pellet machine's decal reads, "Pull out slide, drop in penny,push. Crystalets, a pure delicious confection and a delightful lasting breath perfume. Guaranteed under food & drugs Act June 30, 1906." **Crystal Vending Co.** c. 1908

SPRUCE GUM A full-size stick of gum was dispensed by this 1-column vendor. Around 1900 there were some strange flavors of gun and Spruce Gum came from the resin of the spruce tree. The machine has a wood case with a bronze front. **Curtis & Son** pat. 1894

DEAN A jiggle-slide type of mechanism was used in this aluminum bulk vendor. Some machines have "Dean" embossed on the lid and some are not painted while others are painted bright colors. c. 1960's

DESCO Note that this machine was made in Holland and how much it resembles our Silver King. **Desco** c. 1950

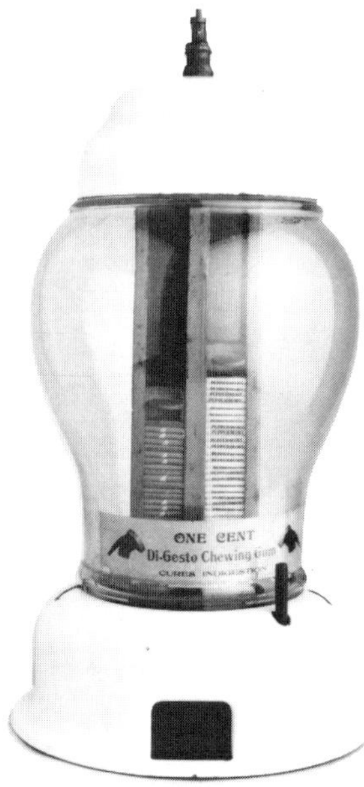

DI-GESTO The decal says "one cent Digesto Chewing gum cures indigestion." The base and lid of the 2-column vendor are made of porcelain. It has been rumored that Pulver was an officer in the company and broke away to form the Pulver Chemical Co. **Di-Gesto Gum Co.** c. 1898

ADAMS' GUM VENDOR This 4 column tab gum vendor was also sold by the G. V. Corp. (Gum Vending Corp.) and the Stewart & McGuire Co., however, it was supposedly made by Du Greiner. It is usually seen with the Adams' decal on the glass. **Arthur H. Du Greiner, Inc.** 1934

ELDREDGE This booth machine was great for cafes and bars as it was both a nut dispenser and a napkin holder all in one. **Eldredge Automatic Mfg. Co.** c. 1940

LITTLE GIANT This early cast iron gumball machine has one of the greatest coin entries I have ever seen. Embossed on the lid is "Little Giant." **Enterprise Pattern Works** c. 1920's

THE PERFECTION Filagree adorns this cast iron and sheet metal peanut vendor. Painted on the body is "1st Put Penny in opening, 2nd Turn Knob, 3rd Push Button." The patent number, 693,157, and company address are embossed on the cast iron lid. **Enterprise Mfg. Co** 1902

ENTERPRISE PEANUTS The Enterprise peanut machine has a clockwork mechanism that vends boxes of spanish peanuts. The case is made of oak. The window is meant to display the peanuts. **Enterprise Vending Machine Co.** c. 1905

CHARMY A squiggly paint design decorates this cast iron and sheet metal machine. The globe is plastic and the lid is chrome. The machine dispensed gumballs and charms. **S. Eppy** 1958

FAUST MAIER Filling this peanut machine had to have been nearly impossible. This machine was also private labeled and was advertised in a Mill's Novelty Co. catalog. **Faust Maier Co.** c. 1908

BREATH PELLET The harp shaped die–cut on the top of this cast iron machine is a marquis holder for an advertising card. **Federal Vending Co.** pat. Jan. 8, 1907

RADIO VENDER The Radio Vender is an aluminum hot nut machine. It has a window to show the customer if there are nuts or not. It utilizes a pull chain to work the mechanism. **Femachen Mfg.** c. 1930's

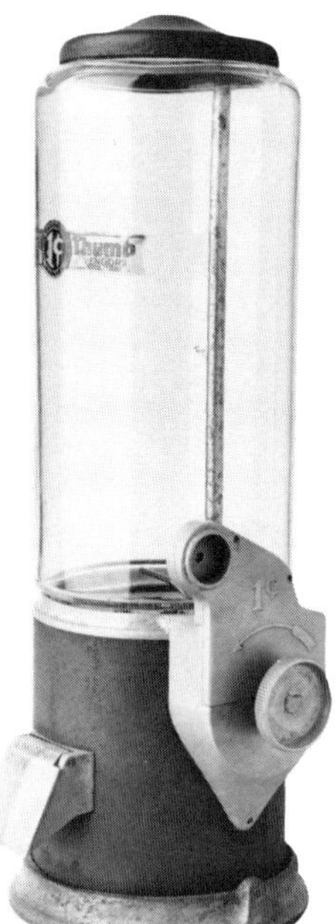

TOM THUMB Walker – Freeman manufactured an earlier version of this machine. An ad for this new version appeared in the January 1936 issue of *Automatic Age* under the Fielding Mfg. Co. name and Walker – Freeman ads ceased. Both companies were located in Jackson, Mich. The machine on the left shows the more common one. **Fielding Mfg. Co.** 1936

FLEUR DE LIS This highly embossed cast iron, 2–column vendor was manufactured for Hopkins & Kost Co. Ltd of Newark, N.J. The front advertising is in gold leaf reverse on glass. **Fleur De Lis Pepsin Chewing Gum** c. 1910

FORD Note how this early Ford has a round globe boldly embossed with the Ford name in script. The early Fords have a dish and the formed metal body was usually painted red. **Ford Vending Machine Co.** pat. 1919

FORD This early Ford has the round globe and the dish in the base, however, what makes this machine special is the great slug rejecter. **Ford Vending Machine Co.**

FORD Ford changed their name from the Ford Vending Machine Co. to the Ford Gum and Machine Co. in 1934. The red formed metal body has decals on the sides and the dish. **Ford Gum and Machine Co.** c. 1935

FORD This early Ford does not have the dish on the base. It has a gumball spout with a cover over it. The rest of the machine remains the same. **Ford Gum and Machine Co.** c. late 1930's

FORD Ford moved their operations from Lockport, N. Y. to Akron, N. Y. This formed metal body with a chrome finish is the model that most of us are familiar with. **Ford Gum and Machine Co., Inc.** c. 1950

FORTUNE GUM The decal on the globe shows a gypsy–type fortune teller and reads "your fortune in each ball of gum". **Fortune Enterprises, Inc.** c. 1930's

FREEPORT TWINS This machine consists of two cast iron fronts on a single wooden box. Each side held 10 lbs., one side for candy and the other for peanuts. Many of the old advertisements promised "blue sky operations." An ad in an old Albert Pick catalog states, "This machine will empty itself on any big day." The machine is stenciled on the back. **Freeport Novelty Co.** c. 1905

FREEPORT DRAGONS The magnificent dragons that adorn the front of this machine are copper plated. I have seen two versions, one with with windows on the sides and this one with ornate castings where the other has windows. **Freeport Novelty Co.** 1905

FREEPORT PEANUT VENDOR An old ad promises, "Everybody has pennies. The Freeport automatically supplies the peanuts for the pennies. Requires little attention—put in peanuts and take out money." The machine displays the last three coins that were inserted. **Freeport Novelty Co.** c. 1905

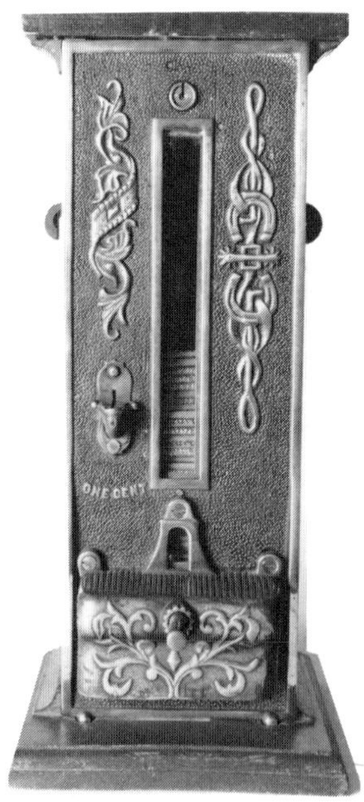

FREEPORT GUM VENDORS Most people call these machine the Goo Goo Gum machine because "chew Goo Goo Gum – good for the gums" is stenciled on the sides. These machines are stenciled "Freeport Gum Vendor" on the back. Both have ornately embossed cast iron fronts and wooden cases. There are two versions, each with a different front casting. One has a long front window that shows off the gum and the other has a short window. Also, one has a protruding coin entry while the other has a flat or flush coin entry. **Freeport Novelty Co.** c. 1910

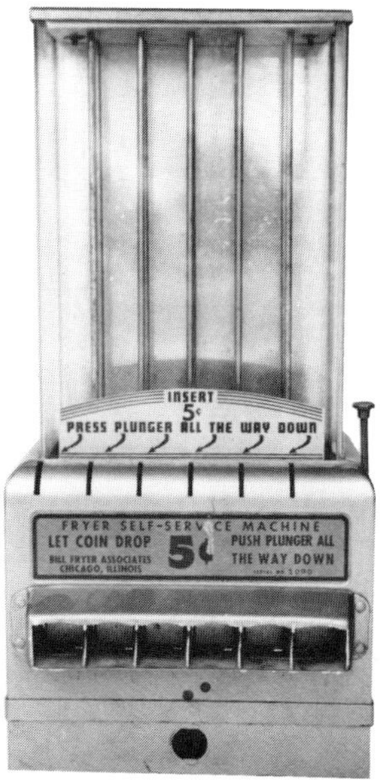

FRYER SELF–SERVICE MACHINE Chicago Gum and Candy Co., National Self–Service Machine, and Gem are the other names that I have seen on this machine. It is a 6–column vendor that dispensed nickel packages of mints or gum. **Bill Fryer Associates** pat. Mar. 26, 1925

GLOBE I have seen some of these machines with "Globe" embossed on the lid and some without. All seem to have "Globe" embossed on the flap. Some people think the machine was made by Simpson because of the coin entry, flap, handle, and round globe. What do you think?

JIM DANDY The base, handle, and lid look like they were made by Advance. A brand of peanuts called Jim Dandies made by the Bordo Products, Co. was also located in Chicago. Perhaps the Jim Dandy was made to dispense Jim Dandies. The large coin entry is also a slug rejecter. **Goober Mfg. Co.** c 1910

U. G. GRANDBOIS The easiest way to identify a Grandbois is to turn it upside down—most of them are marked. Grandbois made both a tall and a short base, available in cast iron and/or cast aluminum. The cylindrical globe came in a 6" or 8" size with lugs on the top and bottom to prevent it from turning on the machine. If a globe is shorter than 6", the odds are the globe has been cut down by the operator because of chips. The Grandbois was available with three different coin entries. The most common was the coin slot entry, as the first three machines show. The second type of coin entry had a push in coin slide like the ones you find on a pinball machine (not pictured here). The rarest coin entry is the large cast iron gooseneck version. The parts are interchangeable with other Grandbois machines as well as with the D. D. Lewis machine. There is a definite connection between Grandbois and D. D. Lewis. Both were located in Kalamazoo, Michigan. **U. G. Grandbois** pat. Sept. 20, 1927

HONEY BREATH BALLS This has to be the simplest mechanism of all time—there are no moving parts! The penny goes down a track and left. The top of the penny knocks the breath ball off the chute and out the machine, dispensing four balls each time, **Gravity Vending Machine Co.** pat. Dec. 1, 1903

GUMBALL I believe that this machine is one of the earliest gumball vendors. When gumballs were introduced in 1907 there were no machines available to vend them. With little alteration, the Honey Breath Ball machine could be made to do the job. Both machines pictured came from Wagner & Miller, Sole Agents. **Gravity Vending Machine Co.** c. 1907

BERKSHIRE The base and lid on the Berkshire are shaped like the Sel–Mor Model E, however, they have raised geometric panels that are plated in chromium. The cylindrical globe is the same as the Model E. **Great States Mfg. Co.** 1938

SEL–MOR The Model E has the cylindrical globe and the large lid. It was finished in a crinkle paint. For an extra 65 cents you could buy the De Luxe model which was primed in chromium. Its shipping weight is 22 lbs. **Great States Mfg. Co.** pat. Jul. 19, 1938

SEL–MOR The model D has the flared globe and the small top. It was available with either a Model E or a Silver King look–alike base and came in a red or green crinkle finish. **Great States Mfg. Co.** pat. Jul. 19, 1938

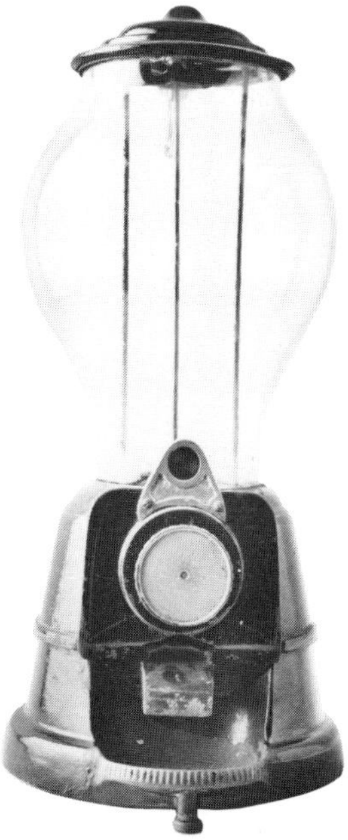

GRISWALD It seems as though Griswald felt that if it was worth making that it should be made in cast iron and this machine is no exception. This model has a round base with the peanut tray built into the base. Embossed on the back is "Griswald Mfg. Co., Rock Island, Ill. **Griswald Mfg. Co.** c. 1914

RED STAR Griswald made this machine for the Red Star Sanitary Vending Machine Co. of Davenport, Iowa. The flared globe holds 7 lbs. of peanuts. The aluminum tray is attached. **Griswald Mfg. Co.** c. 1914

RED CROSS The Red Cross Sanitary Machine Vending Co., for whom Griswald made this machine, and the Red Star Sanitary Vending Machine Co. were both located in Davenport, Iowa. probably indicating a connection between the two companies. The tray is cast into the base. **Griswald Mfg. Co.** c. 1914

SHORT'S CHEWING GUM Long sticks of gum rather than tab gum were vended from this 2–column, cast iron and sheet metal machine. **Griswald Mfg. Co.** c. 1920's

MINT VENDER The body is made out of sheet metal and has three domed columns to hold the mints. The 1¢ denomination dates the machine. **Guaranteed Products Co.** c. 1930's

H & W The Tag on the machine reads, "H & W Machine Co., Inc., John Horn, Abilene, Texas." A gutsy move considering that this machine is a definite Ford ripoff. **H & W Machine Co.** c. 1930's

MICKEY MOUSE & HIS PALS This black porcelain machine was designed with a slug eliminator but Disney stopped production of the machine before it was shipped because the slug eliminator would make the machine too costly. The machine was marketed without a slug eliminator. A 6–color "Mickey Mouse and His Pals" decal goes completely around the globe. Hamilton offered Disney character charms made of plastic wood, but only to operators of the machine. **Hamilton Enterprises, Inc.** 1938

HAMILTON 3 WAY VENDOR Gum – confections – nuts. This was Hamilton's second machine. It had a green porcelain body and a chrome lid. The machine has the same "Mickey Mouse and His Pals" decal as the black machine. **Hamilton Enterprises, Inc.** 1939

REX This cast iron machine is often called the "Rex" because some of the gooseneck coin entries are marked "Rex Mfg. Co. Cols, O." My research in Columbus showed that there was a vending company called the Rex Mfg. Co in 1908 and 1909 and then they disappeared. I believe that the machines with the marked coin entry were made for the Rex Mfg. Co. in 1908 and 1909 and that the machines without anything on the coin entry were for general distribution by Hance. **Hance Foundry Co.** 1908

ZENITH Until recently, there was little known about this machine. It had the "look and feel" of a Hance "White Base" machine but there was no documentation to support this. After seeing a picture, John Hance said that their company made this machine and a cast iron version. **Hance Mfg. Co.** c. 1915

STANDARD REX Hance started making vending machines in 1908 and made them until 1930. In their ads, Hance refers to these machines as the Standard Rex Vendors. They are both part of Hance's "White Base Line." The machine on the left is a penny bulk vendor while the machine on the right with the raised coin entry, is a nickel machine. The older machines had cast iron mid–sections and lids, substituted by aluminum in the newer versions. It has an 8 lb. globe and an adjustable vending mechanism. They also made an all cast iron version. One of their ads reads "Three billion and seven hundred and fifty million pennies made each year. Are you getting your share?" **Hance**

HANCE This is a cast iron bulk vendor that is similar in appearance to the Standard Rex. The machine was available with or without the attachable feet. It was also available in a crackle–paint finish. **Hance Mfg. Co.**

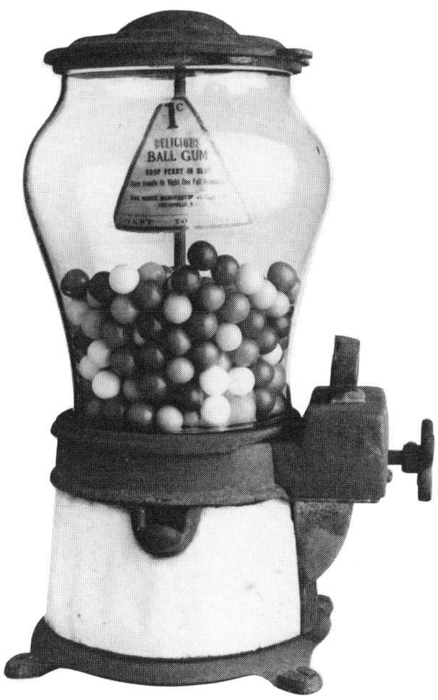

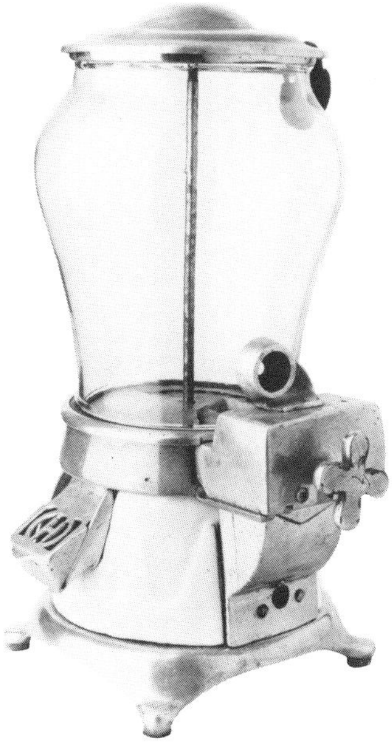

REX GUMBALL I believe that the Rex Gumball is the forerunner of the Rex Convertible. This machine was strictly for gumballs and could not be converted to another machine in the "White Base Line." The base is white porcelain and the rest of the machine is cast iron. **Hance Mfg. Co.**

REX CONVERTIBLE The Rex Convertible, a part of the "White Base" line had an changeable vending wheel allowing the vending of bulk merchandise or gumballs. It had a capacity of 600 gumballs or 4 lbs. of peanuts. A double capacity globe was available for an additional charge. **Hance Mfg. Co**

HANCE There are a lot of minor variations on this machine. The coin entry can be the periscope type as is pictured or a flat coin entry. This one has a single hold–down rod while some have three rods. There are also differences in the feet and the lids. This is a cast iron bulk vendor. **Hance Mfg. Co.**

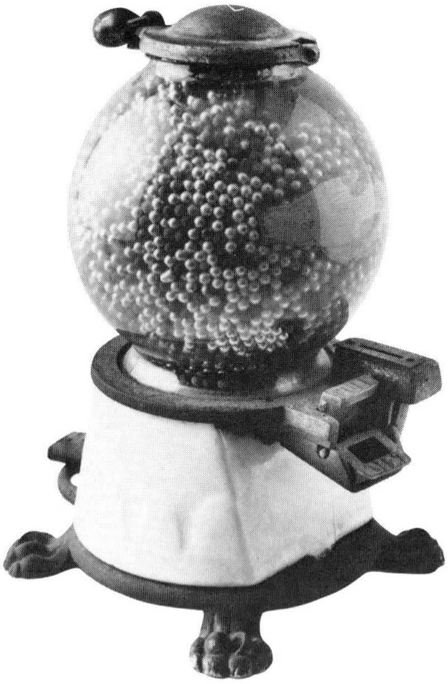

REX BREATH PELLET This little beauty is also part of the "White Base" line. The rest of the machine is nickel plated cast iron. The globe will hold 2 lbs. of breath pellets that cost the operator about $3 and gross him about $14—a nice profit margin. **Hance Mfg. Co.**

REX BREATH PELLET This machine is made completely of aluminum. There were two versions of the aluminum machine. With a profit margin of around 400% one would think there would be more of these machines around. **Hance Mfg. Co.**

HANCE This is another great cast iron bulk vendor. Hance used this flared globe with several different models. Note the pull handle on this model. It was available with and without the attachable feet. I have also seen a hot nut version. **Hance Mfg. Co.**

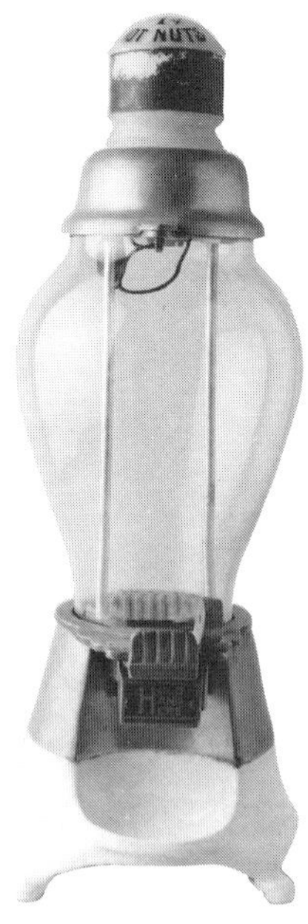

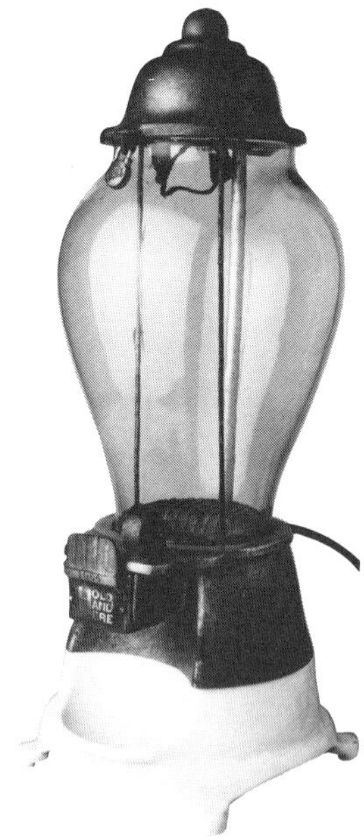

HANCE This cast iron bulk vendor utilizes a pull ring instead of a handle. It has a periscope coin entry and the feet are built–in. The flared globe is the only correct one for this machine. **Hance Mfg. Co.**

REX ELECTRICAL VENDER This machine is the same as the Standard Rex Vender except for the added flash attachment and the heating element. The heating element was made for Hance by Westinghouse. According to Hance's literature, it was very economical to run—only a penny and a half for twelve hours. I have seen this machine with several different tops. On the one that is shown, a green band around the bulb tells you that it is a penny machine, while a blue band stands for a nickel. It was also available with an exposed red bulb. **Hance Mfg. Co.**

H. K. HART This machine is completely made of plastic except for the front mechanism and the bottom plate. Hart used free, all–plastic machines as a sales gimmick to buy their merchandise.

HART This machine is a direct copy of the Ford machine and was meant to compete with Ford. The Hart Gum Company was sued by the Ford Gum and Machine Company for patent infringement. **Hart Gum Co.** c. 1940's

HAWKEYE This cheaply constructed machine is usually found in poor condition. Hawkeye Novelty is still in business under the name of Fawn Engineering. F. A. Wittern, the founder, was reputed to be the maker of the first vending machine to return the correct change. **Hawkeye Novelty Co.** c. 1940's

HAWKEYE As the decal shows, after every ninth pull the customer was rewarded with a free tenth pull. Most of the machines are found without this feature because it was outlawed as a gambling machine. Shown here are the aluminum and the porcelain models. They have two coin slots, the lower being a bent coin or slug return. **Hawkeye Novelty Co.** c. 1930's

HERSHEY'S 1¢ To overcome the fact that the customer was unable to see the merchandise in this 1–column Hershey vendor, the machine returned the customer's coin when it was empty.

HILO ROLL GUM The Hilo Gum Co. copyrighted their name in 1904. Note some of the flavors marked on the front of the machine: Blood Orange; Pepsin Licorice; and Cinnamon. Hilo also made a larger 2–column model. **Hilo Gum Co.** c. 1905

CLIMAX I believe that this machine is the first Hilo machine, as the other Hilo machine is marked "improved" and that it was made for the Hilo Gum Co. by Advance. It has the same patent dates as the Advance Climax: Mar. 6, 1904, Oct. 23, 1900 and the improved Hilo was patented by B. M. Davis, who was a co–founder of Advance. He also had several of the patents on the Advance Climax. **Hilo Gum Co.** pat. Mar. 6, 1904

IMPROVED HILO The patent for this cast iron or aluminum machine was assigned to the Hilo Gum Co. by B. M. Davis, a co–founder of Advance. Since Advance was the likely manufacturer of the Hilo Climax, it is easy to assume that they made the Improved Hilo, too. This model always says "Improved Hilo" on it, so probably followed the Climax model. It came with and without a tray. **Hilo Gum Co.** pat. April 7, 1908

ROYAL ACE This porcelain bulk vendor with a diamond faceted globe was first made by the Baron Sales Co. and was called the Baron Royal Ace. By 1936 the company had become the Hladky Mfg. Co. and the machine, now finished in baked enamel, was called the Royal Ace. **Hladky Mfg. Co.** pat. Jan. 24, 1933

WRIGLEY'S There are three variations of the glass topped Hoff Vendor. Two models are for Wrigley's gum and one is for 1¢ Lifesavers. There are two variations of the aluminum topped Hoff Vendor. The Wrigley's Juicy Fruit is shown here and the other is embossed with Wrigley's PK gum. "Produced for Hoff Vending Corp., N.Y.C. by Doehler Casting Co. Brooklyn, N.Y." is embossed on the machines. They had a pot metal mechanism and very few are found in good working order. **Hoff Vending Corp.** c. 1927

BUTTER–KIST This 1¢ vendor dispensed hot peanuts. They were usually found on a wooden box next to a peanut roaster. Etched in glass is "They're Hot." **Holcomb & Hoke Mfg. Co.** pat. Dec. 25, 1923

SUGAR BOWL The Sugar Bowl is a 3–compartment vendor. A flat roof, nickel compartments, different chute flaps and different handles are some of the variations that I have seen. **Holli–Ware Mfg. Co.**

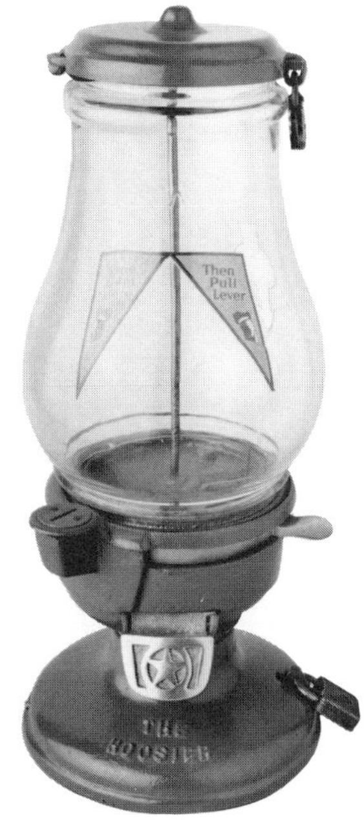

HOOSIER This machine looks very similar to a Columbus Model A, however, not all of the castings are interchangeable. Not having seen any but this one, it is difficult to tell if the lantern type globe is correct. Could this have been made in Indiana or for someone in Indiana?

IMPERIAL Each time the customer pushes the handle to dispense a piece of chocolate this machine also pops the three dice under the dome. If the dice total 3 or 18, the customer won a valuable prize. **Imperial Mfg. Co.** c. 1910

INDUSTRIAL ENTERPRISES Even though this picture shows gum in the machine, it is a nickel bulk merchandiser. It is made of cast iron and has "Industrial Enterprises, Chicago, Il." embossed on the lid. **Industrial Enterprises** c. 1930's

JACOB'S GUM Very few of these pot metal machines have survived. The company's name and the patent date are marked on the bottom. **Jacob's Gum Co., Inc.** pat. July 4, 1916

IN THE BAG These bulk vendors dispense merchandise in a glassine bag. It was supposed to be sanitary since the customers hands do not touch the merchandise. Through the years there were several minor variations including machines with and without marquees and penny or nickel machines. **O. D. Jennings Co.** c. 1934

MINT VENDOR When you insert a nickel in the top of a column, it automatically releases a roll of mints. The three column vendor is made of sheet metal. **O. D. Jennings & Co.** 1932

PLAY BASKETBALL Three shots for a penny! If the price of sugar went up, the operator could give the customer fewer shots for his money. There was also a Play Football version. **E. E. Junior Mfg. Co. Ltd.** 1930

BEECH–NUT CHEWING GUM This machine was available in a Wrigley or a Beech–Nut version. It was interchangeable for either penny stick gum or nickel pack gum. **Kayem Products Co., Inc.** 1947

KEMACO Kemaco is an acronym formed from the first two letters in the words *Kelley Manufacturing Company*. The machine on the left is the earlier version and was made by the Kelley Gum and Mfg. Co. The raised coin entries and the designs around the word *Kemaco* are different on the two machines. The older version has longer pinstriping on either side of the handle while the newer version has embossed filagree and shorter pinstriping on either side. **Kelley Mfg. Co.** pat. Aug. 23, 1904

LION All the Lion vendors that I have seen have a metal tag reading, "The Novelty Cigar Co., 209 Commonwealth Bldg., Portland, Oreg." The Lion vendor is like the later Kemaco. **Kelley Mfg. Co.** Pat. Aug. 23, 1904

PIX The Pix is a private label machine made for Albert Pick & Co., which was a distributor for Kelley. It has an embossed design on the lid, a different design around the name *Pix*, and a different embossing on the front of the base. **Kelley Mfg. Co.** Pat. Aug. 23, 1904

CLIMAX I believe that this machine was made for Kelley Mfg. Co. by Advance. It is the same as the Advance Climax except it has Kelley Mfg. Co. on it. It even has Climax embossed on the machine. Advance patent dates embossed on the machine are: 8–27–07;–7–2–07; 8–20–07; 3–8–04.

BASEBALL To play the game you flipped your gumball to see how many runs your team had scored. The machine is made of sheet metal **Kholm Mfg. Co.** c. 1950's

KOMET This is a newer machine that is usually found promoting some organization or fund. The globe is made out of plastic. c. 1960

SELECTIVE This bright yellow tab gum machine is made of sheet metal. You are able to turn to different columns by turning the knob on top. **Kopper–King** c. 1950's

LAWRENCE All of these 2–compartment bulk vendors that I have seen have been green and chrome. **Lawrence Mfg. Co.** c. 1940's

LEE The coin is pulled from the horizontal coin entry into the machine when you pull the lever. It is found with and without an embossed globe. I found one with the bottom embossed "Valentine Produce." c. 1930

LEEBOLD The Leebold is a heavily ornate Art Nouveau aluminum vendor. The one found most often is pictured on the left. They are embossed on the back, either with "St. Louis" or "San Francisco." The Leebold on the right is identical to the left one except that the coin drops through a chute with a window. These are embossed "Berkeley, Cal." There is strong evidence that R. D. Simpson may have made the Leebold! **C. E. Leebold** c. 1920

ADVERTISING LEEBOLD This model has several differences. It has an extension to the lid with three panels of advertising. "The Leebold" is not impressed on the front and there is no city embossed on the back, lending weight to the argument that Leebold was not the manufacturer. **C. E. Leebold** c. 1920

LEEBOLD Embossing around the coin slot reads, "Put Coin in Slot, Revolve Handle Full Turn." The decal, which is different, reads, "Millions Eat Leebold's Salted Peanuts. There is more nourishment in 5¢ worth of Leebold's Salted Peanuts than in a porterhouse Steak." **C. E. Leebold** c. 1920

D. D. LEWIS The brass tag on the base identifies D. D. Lewis machines. It has a cylindrical globe that is open at the top instead of a domed shaped globe. This cast iron miniature gumball machine was used in locations where space was short. The locks in the top and base are reverse threaded. Note the similarity between a D. D. Lewis and a U. G. Grandbois, also made in Kalamazoo, Mich. The parts are interchangeable. There is a definite link between the two companies. **D. D. Lewis Co.** c. 1920's

LIBERTY GUM This machine was a cheaply constructed, pressed metal gumball machine. Note how primitive the handle is. c. 1915

LINCOLN I believe that the Lincoln was made by the R. D. Simpson Co. It looks like the Simpson with claw feet and it is finished in chrome, which is something Simpson liked to do. Three stained glass panels on top of the machine are lit by a light bulb. I have also seen ads for a cold nut Lincoln. c. 1915

LITTLE NUT Also known as the Model B, this first model was a miniature table vendor made of cast iron and aluminum. Coin entries can vary. **Little Nut Vendor Co.** 1934

NUT HOUSE This machine, shaped like a house, was Little Nut Vendor's second entry into the vending industry. It is made of cast iron and sheet metal. **Little Nut Vendor Co.** 1935

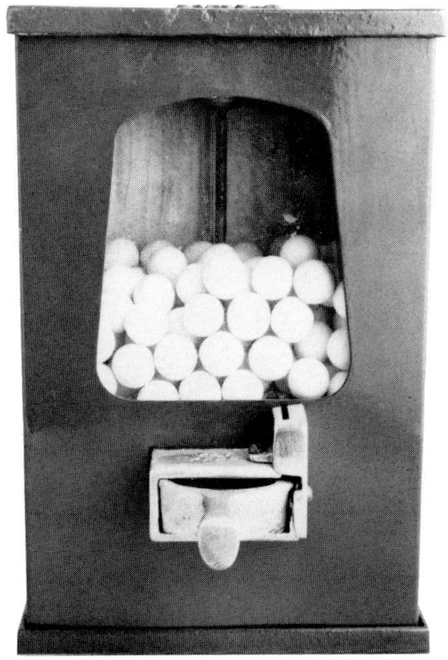

NUT SHOP This is Little Nut Vendor's third entry into the business and their third with a jiggle–slide type of product delivery. It is made of sheet metal. **Little Nut Vendor Co.** 1935

TYPE C This highly polished cast aluminum machine supposedly would vend anything: peanuts; pistachios; small candies; and gumballs. **Little Nut Vendor Co.** 1935

TWIN VENDOR This machine is a twin Type C vendor. Available in a counter model or a standard floor model that sold for the same price as two of the Type C Little Nuts. **Little Nut Vendor Co.** 1935

TYPE C The base on this machine is made out of cast iron and has a rim along the bottom. It is finished in black crinkle paint and has a large globe. **Little Nut Vendor Co.** 1935

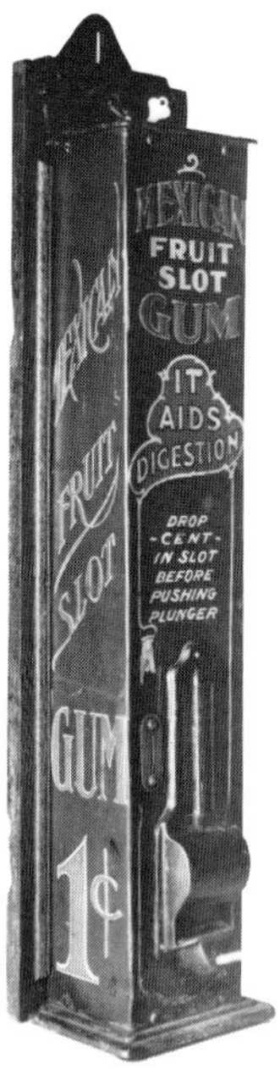

MEXICAN FRUIT The embossed, lithographed tin sides of this machine advertise Mexican Fruit Slot Gum and how it aids digestion. London Novelty Company also made a Pansy Gum machine. **London Novelty Co.** c. 1900

PANSY GUM As the machine says "a big stick of Pansy Gum, your fortune, and a love letter all for 1¢." A lot of the lithography on the case is embossed. Note the similarity to the Mexican Fruit Slot Gum Machine which was also made by the **London Novelty Co.** c. 1900

SUN This is a cast aluminum 5¢ nut machine. Some are embossed with the word "Sun" on top and some are not. Some have Los Angles Mfg. indented in the flap and some don't. It uses a hex wrench for a key. **Los Angles Mfg. Co.**

I SELL EM HOT (left) **EAT EM HOT** (right) Determining the proper manufacturer is a confusing issue. Some of the machines are marked Mabey Electric and Mfg. Co., some are marked Roy Stringer Co., and others Cebco. As Mabey and Roy Stringer used the same addresses in their ads, one can assume they were the same company. Both machines were patented to Charles Mabey. I Sell Em Hot was patented Oct. 24, 1922 and Eat Em Hot was patented in 1934. **Mabey Electric & Mfg. Co.**

BAKER BOY (left) **BLACK MAN** (right) When the customer cranks the handle, the Baker turns to the oven, scoops out a gumball, nods and drops the gumball down the chute. The Baker Boy is the most frequently found model. The Black Man is a different automaton. It works the same way as the Baker Boy. **Maniken Vendor Co.** pat. Aug. 27, 1929

BELL HOP (left) **INDIAN** (right) The Bell Hop and the Indian are two more automated characters that were found in the Maniken vendor. All the automatons go through the same gyrations. **Maniken Vendor Co.** pat. Aug. 27, 1929

LOOP THE LOOP Diavolo (note the different spelling) was the man who first "looped the loop" at Madison Square Garden according the *The Pictorial History of American Circus.* The Loop the Loop was shown in the 1904 Caille Bros. catalogue and it was assumed that Caille was the manufacturer of the machine, however, the patent, decals, original mechanical directions, and shipping stickers support that the Mercer Mfg. Co. was the maker. This means that it is likely that Caille was a distributor. **Mercer Mfg. Co.** Pat. June 28, 1904

MATTHEWMAN This machine is a prototype that never got off the ground. To operate the machine you turned a ring at the top of the machine to the month of your birth and then turned another dial to the day of the month. For a nickel, the machine dispensed a package of gum and an astrological prediction. **Matthewman Mfg. & Specialty Co.** c. 1930's

MERCER PEANUT This wooden vendor dispensed one cent packages of peanuts. The price of peanuts tells you how old the machine is. This is the same company who made the Loop the Loop. **Mercer Novelty Co.** c. 1905

STRIKE IT RICH The coin falls through a coin drop field giving the customer a chance to receive his coin back after getting his merchandise. This machine uses a see–through plastic coin drop that allows the customer to view the merchandise as well as the coin drop field. **Metalsmiths** c. 1950's

TASTY HOT NUT VENDOR The heating unit for this cast iron machine is a 25 watt lamp. A red light flasher is located in the base below the handle. Options for this machine included electricity, a lid with a flashing light or a flat lid with no light. Ideal, shown on the decal, is a distributor. **Metcraft Corp.** c. 1930

7–WAY GUM VENDOR The 7–Way sold nickel packages of gum. Each row held 20 packages of gum. Made of polished aluminum and nickel plate. **Wm. Michael Co.** c. 1934

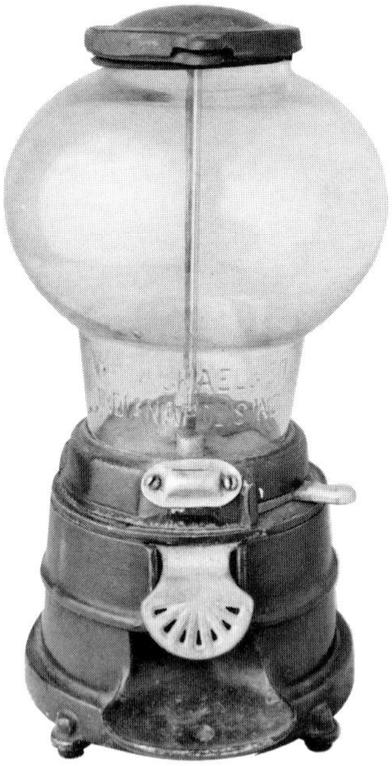

WILLIAM MICHAEL PEANUT The base of this machine is exactly like the National peanut base. "Wm. Michael & Co., Indianapolis" is embossed on the globe. National probably manufactured this machine. **William Michael Co.** c. 1930's

WILLIAM MICHAEL GUM Wm. Michael put out a profit sharing 1–2–3 as well as the regular gumball model. Many people think that the National Novelty Co. made the gumball machine for Wm. Michael, as the machines are so similar. **William Michael Co.** c. 1930

MIDGET VENDOR Another miniature vendor with a jiggle slide type mech. Even though the machine has a cast iron base, it still looks top heavy to me. Embossed on the lid is Midget Vender Co. **Midget Vender Co.** c. 1930's

MIDGET VENDOR The similarities between this miniature machine and the other Midget Vendor end with the jiggle slide mechanism. "Midget Vending" is embossed on the base above the jiggle slide. **Midget Vending** c. 1930's

MIKE'S VENDOR With the repeal of prohibition in 1933 came a flood of bar type vendors. Nuts and beer go great together. This is another miniature machine with a jiggle–type mechanism Embossed on the lid is **Mike's Vendor.**

MILES–HI GUM This porcelain machine weighs in at a mere 34 lbs. and the advertising is in the porcelain. The globe is embossed "Miles Autostack Vending Co., Chicago, pat Mar. 15, 1927." **Miles Autostack Vending Co.**

WILBER–SUCHARD CHOCOLATE Embossed on this cast iron, art deco, 4–column vendor is Wilbur–Suchard Chocolate. The mechanism works the same as the stainless steel Mills' Automatic Tab Gum Machine. The metal tag on the bottom of the machine reads, "L. Miles, Exclusive Dist." **L. Miles**

GUMBALL This Millard Gumball machine is made of tin and is chrome plated. It utilizes the unusual mechanism that is standard for Millards. It was used as a promotional machine.The machine was promoted as a $7.50 value and was sold for $6.75 including 1000 gumballs. **Millard Vending** Aug. 29, 1916

BREATH PELLET Millard also made machines for distributors. The $5 price of this cast iron machine included 3 lbs. of breath pellets. The coin drops through a chute in the center of the machine into a mechanism in the base. Turning the top knob vended the merchandise. **Millard Vending** pat. Aug. 29, 1916

GUMBALL Amazingly, both the base and top of this machine are made of wood. The machine looks and works like other Millard machines despite its wooden construction. The decal states, "Deposit Penny on Top, Turn Knob to Right." **Millard Vending** pat. Aug. 29, 1916

MILL'S AUTOMATIC This 6–column, stainless steel vendor was patented to B. E. Mill's. I don't know if there is any relationship between this machine and Wilbur–Suchard Machine Dist. by L. Miles. Both have the same unusual mechanism. **Mill's Automatic Merchandising Corp.** pat. 1936

ADAMS' GUM This L–shaped vendor is unusual in that it is painted instead of having porcelain panels. **Mill's Automatic Merchandising Corp.**

OWL GUM This machine could be adapted to vend gum, matches, or chocolate; making it very versatile. The machine not only came as a single but was also available as a double or triple. **Mill's Novelty Co.** c. 1905

MILL'S GUM VENDOR Some people say this machine was made by the Mill's Novelty Co. because their name is stenciled on the back and others say the machine is a Goo Goo because Goo Goo Gum advertising is stenciled on the sides. Actually the machine was made by the Freeport Novelty Co. of Freeport, Ill. and was distributed by Mill's. The machine dispensed stick gum. **Mill's Novelty Co.** c. 1899

LITTLE PERFECTION Mill's ads told how nicely the large clear glass globe displayed the peanuts and how easy it was to fill through the top. The machine was supposed to be water and dirt–proof. With the sanitary cup it was even supposed to be dust–proof. It was constructed of cast iron and wood. **Mill's Novelty Co.** pat. Sept. 10, 1912

ORBIT Note how similar the Orbit looks to the Mill's Imperial Shocker. This baby was no lightweight as it weighed in at 70 lbs. No wonder they bragged about it being substantially built. This cast iron beauty was made to dispense peanuts or small candies and was supposed to be slug–proof. This is one of the earliest peanut machines. **Mill's Novelty Co.** 1902

ORBIT JR. From the pictures I had seen and the name Orbit Jr., I had always visualized the machine to be a small breath pellet machine. True, it is meant to vend small candies or peanuts, but I sure guessed wrong on the size. The machine stands 22" and weighs in at a mere 50 lbs. Like its sister machine, the Orbit, it is a very ornate cast iron bulk vendor. Note the tin advertising signs around its perimeter. It existed in the same time frame as the Orbit. **Mill's Novelty Co.** c. 1902

MODERN I believe that the Modern was made by the Universal Sales Machine Co. who also made machines for Remington Service Machines, Inc., Unxld Gum Co., and the National Gum Co., Inc. It vends 6 gumballs for 5¢ and 13 balls for 10¢. **Modern Vending Machine Co.** c. 1920's

MO–JO It is amazing to me that any of these domed vendors survived at all. The globe is put on upside down and held on by a steel tension rod. The side reads "Put Finger in Loop. Push Plunger." **Mo–Jo** c. 1915

NATIONAL The National was a private labeled machine made by Oak Mfg. The machine has been altered by putting a Liberty Bell on the flap instead of an Acorn. **National** c. 1950's

SWEET CHOCOLATE The Model S sold 1¢ tablet chocolate and the Model G sold 1¢ tab gum. Both are of sheet metal construction. I have also seen ads for these machines under the name of the National Gum Co., Inc. of Newark, N.J. **National Automatic Vending Machine Co.** pat. Dec. 18,1917

LITTLE KING This machine is cast aluminum with the exception of a cast iron base. It has a jiggle slide type of mechanism. **National Coin Machine Exchange** c. 1935

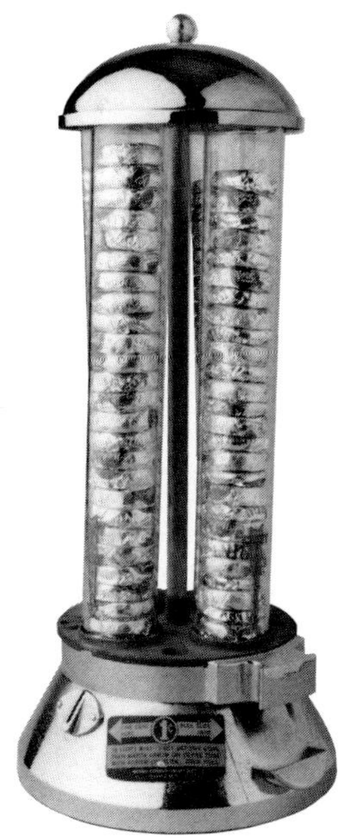

FOUR–WAY National Dispenser Co. was the sole distributor for the Four–Way Selective Mint Dispenser that dispensed 1¢ candy mints. The four cylindrical glass columns rotated lazy–susan style. **National Dispenser Co.** pre 1939

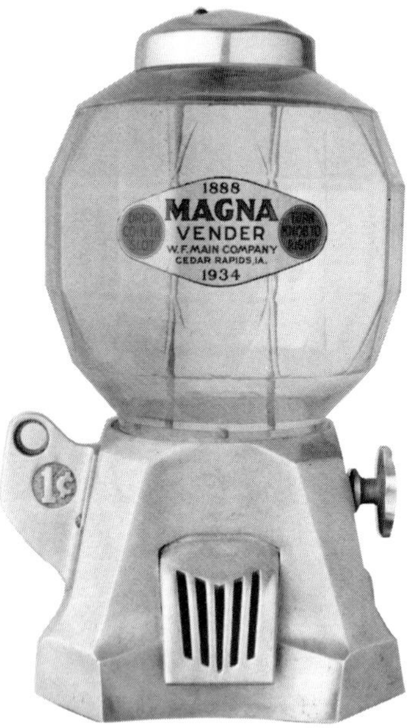

MAGNA VENDOR This machine was patented by W. F. Main and his name is on the decal. Actual manufacturing was done by the National Mfg. Co., the company who made the Double Nugget. The globe is 16–sided and the base is 8–sided. There were penny and nickel models. **National Mfg. Co.** pat. July 31, 1934

DOUBLE NUGGET W. F. Main patented this polished aluminum double bulk vendor. Some chute flaps have vertical slits cut in them and some have the letter "N" stencil–cut in them. National also made the Magna Vendor. **National Mfg. Co.** pat. Feb. 2, 1937

BREATH PELLET The decal reads "Fragrant Wholesome and Lasting Breath Perfume." A bulbous glass globe sits on a miniature 8–sided body with claw feet. **National Novelty Co.** c. 1910

NATIONAL PEANUT It is unusual to find a wooden peanut machine. The machine had a metal container inside to hold the nuts. A large beveled window in the front displayed the nuts. **National Novelty Co.** c. 1910

NATIONAL NOVELTY CO. This cast iron peanut vendor has a tear–shaped globe. It is believed that National also made the William Michaels peanut vendor. **National Novelty Co.** c. 1910

NATIONAL The national was a gumball machine which either was made out of porcelain or had a painted, cast iron body with a formed metal lid. It is presumed that they also made the William Michaels gumball machine. **National Novelty Co.** c. 1910

BLACK DIAMOND PEPSIN GUM The sides read, "We chew Black Diamond Pepsin Gum. We chew it 'cause most others are bum. We chew it now and want some more to chew, until we can chew no more." Bethlehem Mfg. Co. made the gum. **National Vending Machine Co.** pat. Mar. 21, 1905

NEKO BULK VENDOR The design of this bulk vendor has art deco influences. The top and bottom are made of cast iron and the body is made of sheet metal. A pull chain operates the mechanism. **Neko** c. 1930's

DOUBLE The Neko Double has a single cast iron base and a cast aluminum top. The green finish is decorated with gold, squiggle, dribble–paint. A pull chain operates the mechanism. **Neko** c. 1930's

MASTER STAR DOOR The early Masters were nicknamed the Star Doors for obvious reasons. Apparently Norris hoped to capitalize on Columbus Vending Co.'s popularity by copying the circled star. I would guess that Columbus balked about this as the Star Door was short–lived and was only found on early Masters. This machine does not have the name "The Master" on it nor does it show a patent date and only reads "Patent Applied For." The turning knob is cross–shaped instead of flat. The machine has an aluminum slide top to match the aluminum cabinet. The cabinet does not have a center rod. Below the chute flap is an indented tray. The coin door has a filagree design instead of the rising sun design. It was available in a penny, penny–nickel, or a 1–2–3 version. **Norris Mfg. Co.** c. 1922

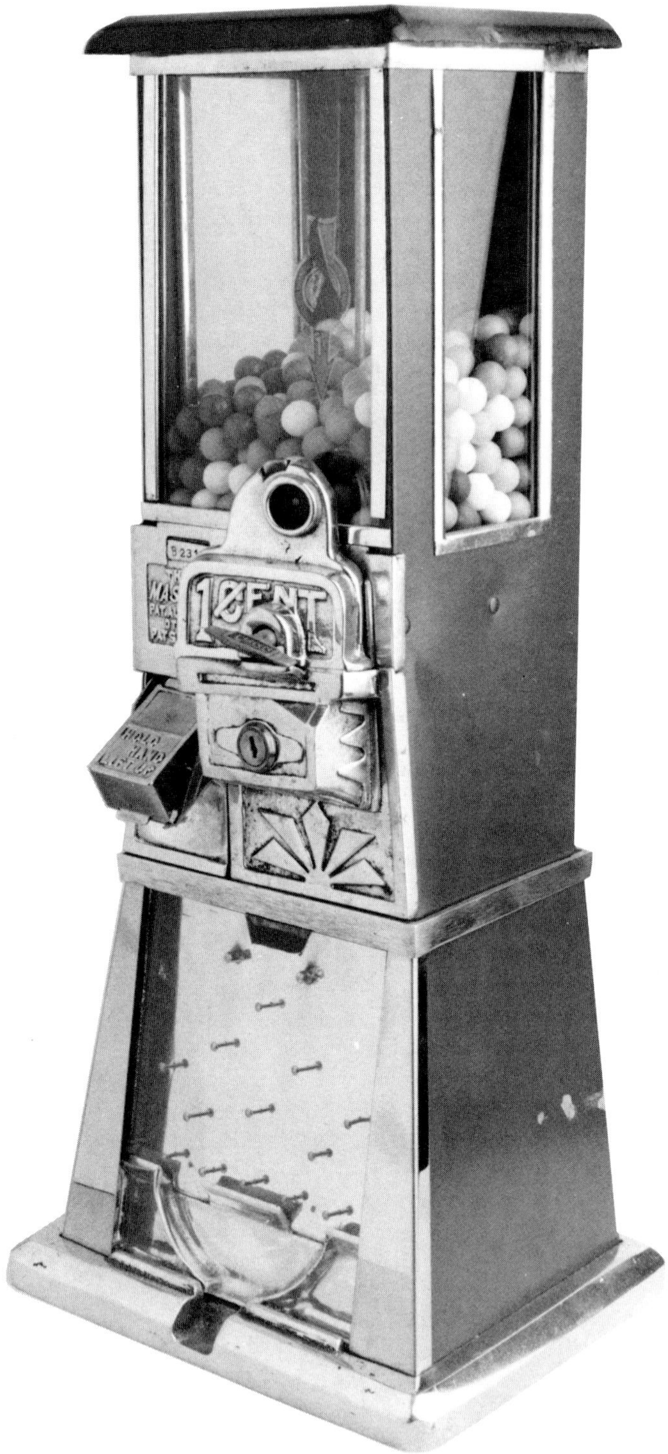

MASTER HOT NUT This hot nut machine is a Master #2 with a raised lid that slides on. The milk glass panels in the lid light up. As the machine is made out of aluminum and has a sliding lid, it suggests that this is an early Master. **Norris Mfg. Co.**

MASTER PENNY DROP This machine is a standard Master on top of a tall base with a penny drop front. After going through the Master, the penny would drop down the play field. The customer always got his gumball and stood about a 10% chance of getting his penny back. **Norris Mfg. Co.**

MASTER (aka Master Novelty and Master Special) This, the most common of the Masters, is typically marked "Pat. Aug. 14, 23 Other Pats. Pend." and the chute flaps are marked "Hold Hand, Lift Up." Some customers, such as Lazar, put their own name on the flaps. A rising sun design is usually found on the coin door. Most of the parts were interchangeable. Some were painted steel but most cabinets were porcelain available in red and black,red and white, tan and green; white and green; blue and white; and all white. **Norris Mfg. Co.** pat. Aug. 14, 1923

MASTER NO. 77 Norris calls this machine the No. 77 Penny–Nickel, however, collectors have dubbed the machine the Fantail. There are different slots for pennies and nickels, one turn for a penny and five turns for a nickel. Rarer versions are the straight penny Model 77P and the nickel Model 77N. Four slug detectors are housed in the fantail: a magnet to catch ferrous slugs; a hook to catch washers; a weighing device that returns improperly weighted slugs; and a tilt mechanism that returns coins unless the machine is level. **Norris Mfg. Co.** 1933

MASTER NO. 2 Collectors refer to these machines as gooseneck Masters. The ornate gooseneck houses two slug detectors, a magnet to catch ferrous slugs and the other to catch slugs made out of paper or cardboard. You get one turn for a penny and five turns for a nickel. It was available in a porcelain cabinet or in an aluminum cabinet with a porcelain base and lid. These machines are usually found with the Aug. 1923 and Oct. 1924 patent dates embossed on the front and with filagree or a rising sun design on the coin door. The lid is held down by screw type cylinder locks in the center of the lid. Available versions included penny, penny–nickel, and nickel play. What is unusual about the nickel version is that it only gives one turn for a nickel. **Norris Mfg. Co.** c. 1925

PACKAGE GUM MERCHANDISER Northwestern's first gum machine had four selections of stick gum and was made of white porcelain enamel. The customer turns the revolving magazine to his choice, inserts a nickel and presses the plunger to get his package of gum. **Northwestern Corp.** c. 1917

PURE–GUM MERCHANDISER The Ball Gum Merchandiser is another name for this cast iron machine finished in white–duco lacquer. Variations include a round knob or a handle-type knob, and gumball chutes on the right or left. **Northwestern Corp.** pat. Oct. 7, 1919

SELLALL MINT and GUM VENDER This cast iron, single–column nickel vendor can distribute either packages of gum or rolls of mints. **Northwestern Corp.** c. 1925

HERSHEY'S HONEY BAR This machine is best known as a match vendor, but with a little modification it could dispense many boxed items such as chocolate, soap, aspirin, or sanitary napkins. It has a cast iron base and top with a sheet metal body. **Northwestern Corp.** c 1925

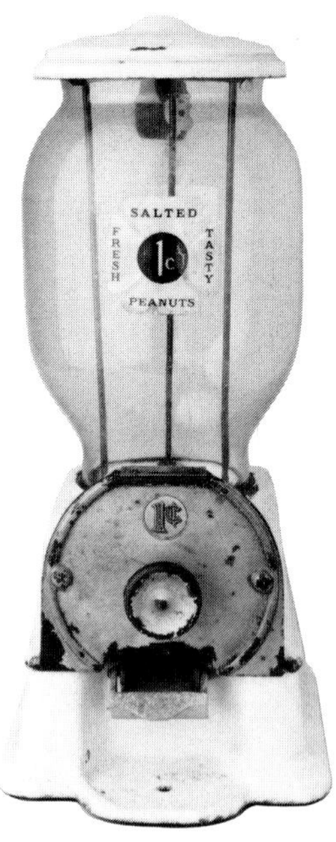

PEANUT MERCHANDISER This was Northwestern's first globed machine and was finished in white porcelain enamel. The machine was heavily advertised for many years. Where did they all go? **Northwestern Corp.** c 1920's

MERCHANDISER First advertised as a mint vendor and then advertised to vend anything, this machine was available in cast iron or porcelain. Initially available with padlocks and after their introduction in 1933, with barrel locks. If the handle is on the right it's a penny mechanism and if the handle is on the left it is usually a penny with a slug eliminator, penny–nickel, or a nickel mechanism. The slug eliminator was introduced in 1932. An optional marquee, on the front, told the customer the contents of the machine. The placement of the chute varied. The machine was also available on a penny drop as is shown with a Model 33. **Northwestern Corp.** 1931

NORTHWESTERN 33 PEANUT This deco–influenced machine was available in cast iron or porcelain, with the vast majority being porcelain. I have seen the following colors: red, green, black, white, yellow, dark blue, and brown. The original 5 lb. globe had "Northwestern" embossed on it and a modernistic design stippled in the glass giving it a frosted effect. After several years the original globe was replaced by a 5 lb. clear cylindrical globe. The $3\frac{1}{2}$ lb. clear cylindrical globe that is on the 33 gum machine also fits and is correct. Available with a barrel lock or a padlock. This machine was only available in a penny version. **Northwestern Corp.** 1933

PENNY DROP This cast iron penny drop was actually made by Northwestern to fit either the Merchandiser, which was introduced in 1931, or the Model 33 peanut machine. The odds were 10% that a customer would get his penny back. **Northwestern Corp.**

NORTHWESTERN 33 JUNIOR This miniature machine was designed to supplement regular vendors in locations where space was limited. The mechanism is identical to the Model 33 peanut machine. It was available with a 1½ or 2 lb. octagon shaped globe. This cast iron machine is found in porcelain or lacquer finish. **Northwestern Corp.** 1933

CANTEEN This is a Northwestern 33 Jr. that was private labeled for Canteen. It has a different chute flap and an added marquee. **Northwestern Corp.** 1933

PROTOTYPE Northwestern agreed that the numbers on the base of this machine are the same ones used on the Northwestern 33 parts. Note the differences in the aluminum base, hold–down rod, and mechanism. The chute flap is missing and I put the 33 globe and lid on the machine. **Northwestern Corp.** 1933

NORTHWESTERN 33 GUM This machine only comes in porcelain, in the same colors as the Model 33 peanut machine. It usually came with the 3½ lb. globe, but the 5 lb. globe, usually found on the peanut machine, also fits and is correct. **Northwestern Corp.** 1933

DE LUXE This ultra–modern looking penny–nickel machine did not seem to stay around very long, probably because of its complicated mechanism. It is the sister machine to the Triselector and is found in red or yellow porcelain. **Northwestern Corp.** 1936

NORTHWESTERN 39 This bulk vendor was only available in porcelain, however, you could pick from a wide variety of colors. Red was, by far, the most prevalent color. It was available with a large or small globe. **Northwestern Corp.** 1939

NORTHWESTERN 39 BELL The 39 Bell was developed to compete with the Hawkeye. It is the same as the Northwestern 39 in every way except that every tenth pull would ring a bell and then the customer would receive a free portion. **Northwestern Corp.** 1939

NORTHWESTERN 40 The Model 40 was an economy version of the Model 39. It was made of pressed steel or porcelain. The Model 40 has a vertical, flush–mounted coin entry instead of the horizontal one on the Model 39. There are not many survivors in good condition. **Northwestern Corp.** 1940

NORTHWESTERN 49 Northwestern's first post war vendor was available as a penny or nickel bulk vendor. It is made out of pressed steel and came in the following colors: red, tan, green, yellow, or white. **Northwestern Corp.** 1949

NORTHWESTERN TAB GUM This machine has the same body and coin mech as the Model 49. The revolving head has 10 selections, 10 columns, and a large capacity of 515 tabs of gum or candy. **Northwestern Corp.** pat. 1954

NORTHWESTERN 5¢ PACKAGE GUM This machine has the same body and coin mechanism as the Model 49. The revolving head has 5 selections, 5 columns, and a total capacity of 95 packs. **Northwestern Corp.** 1954

JET Northwestern made the Jet to compete with the Victor Baby Grand. It was made to vend charms and ball gum or capsules. Available in penny or nickel versions, it had a formed metal body with a curved plastic window. **Northwestern Corp.** c. 1954

NORTHWESTERN 60 This machine was actually made in 1959. It is still being made today. It has been Northwestern's most successful machine with over one million sold. It is sometimes found with the operator's name on the flap. **Northwestern Corp.**

SATURN 2000 The Saturn 2000 holds 2000 gumballs. It is a large machine with a height of 53 inches. The machine was made out of sheet metal and was shaped like a rocket ship. **Northwestern Corp.**

ACORN WITH STAMP DISPENSER A customer would get one portion of merchandise and one stamp for a penny and five portions of merchandise and five stamps for a nickel. A nickel allowed five turns of the handle. The stamps were redeemable for items in premium books. **Oak Mfg. Co., Inc.**

ACORN Embossed on the lid of this machine is "Oak Mfg. Co., Inc. L.A. Cal. Pat. 2,337,317." It has an acorn embossed on the square globe and impressed in the flap. Oak would private label machines for other manufacturers or distributors. Oak was founded in 1947. **Oak Mfg. Co., Inc.**

ACORN This machine has a round lid, a larger cylindrical globe, and an extended base that held more money. **Oak Mfg. Co., Inc.**

OAK PREMIERE The customer received a gumball and a picture card, all for a penny, from this machine. It holds 800 cards and 1000 gumballs. **Oak Mfg. Co.** 1956

ACORN This Acorn is unusual because it has a viewing window that shows the customer what color gumball is next. This machine has the tall rectangular globe on it. **Oak Mfg. Co.**

HOT NUT This is the hot nut version of the Acorn. The machine has been electrified to heat the peanuts. **Oak Mfg. Co.**

VISTA The decal on this machine is incorrect. Even though this is a 1¢ and 5¢ machine, the model is a Vista not an Acorn. The Vista is a later model machine than the Acorn. **Oak Mfg. Co.**

VISTA This machine shows the Vista Model with the plastic globe. **Oak Mfg. Co.**

MODEL 1 (left) **and MODEL 2** (right) This machine was available in polished aluminum or red–enameled cast iron. Model 1 was the peanut version and Model 2 was the gumball version. The base is very similar to the Ace Model 12. **Ohio Vending Mach. Co.** c. 1930

HOT NUT As this is a hot nut machine, it would be a Model 1, which is the peanut version. The light inside the globe was the heater. The attachment on top of the machine is a cup holder. **Ohio Vending Machine Co.** c. 1930

ACE MODEL 43 This polished aluminum machine was the top of the Ace line. It uses the same mech, globe and lid as the model 12, however, the base and the chute flap are different. **Operators Vending Machine Supply Co.** c. 1935

ACE MODEL 12 The Ace Model 12 and Ohio Vending Machine Co. machines are often confused. Only the bases and the mechanisms are similar. The Ace globe is octagonal, embossed "OVMS", but the Ohio globe is round. They also have different lids. The Model 12 was a cheaper model of the Ace Model 34. **Operators Vending Machine Supply Co.**

ACE Art Deco influences are evident in this aluminum bulk vendor. This model is seen less often than the other Ace machines. The Operators Vending Machine Supply Co. later became Acme Vending Co. **Operators Vending Machine Supply Co.** c. 1930's

MITY-MITE I have found two names for this vendor. In the Dec. 1934 issue of *The Coin Machine Journal,* it was advertised as the Dixie Nut Machine made by the Dixie Nut Co. It was finished in bronze and a sample machine sold for $5. In the March 1935 issue of *Coin Machine Review* the machine was advertised as the Mity-Mite, made by the R.H. Osbrink Mfg. Co, which was an aluminum foundry and machine shop. Their sample only cost $3.50. This small nut machine was especially made for bars and booths. Remember they repealed prohibition in 1933. **R. H. Osbrink Mfg. Co.** 1934.

PACK IT SHOP Note the similarity between the Pack It Shop, whose manufacturer is unknown, and the Tid–Bits Shop, which was made by the Venitor Corp. Could Venitor have made them both? Note the stylish art deco globe on the Pack It Shop. **Pack It**

PANSY GUM The customer received a fortune, love letter, and a stick of gum, all for a penny, from this beautiful, ornately embossed 2–column vendor. I have seen this machine made out of aluminum or cast iron. **Pansy Gum** c. 1905

BASEBALL This machine was patented as a vending casino. I love the premise of "double or nothing." The machine drops a steel ball and the player manipulates the baseball player to catch it. If successful, the customer gets two pieces of gum and if he misses—nothing. Paupa & Hochreim was eventually taken over by Mill's Novelty Co. **Paupa & Hochreim** pat. Sep. 3, 1918

PENNY CARNIVAL The coin entry on this machine is a coin slide. Upon insertion of the penny the gumball drops into a pull out tray. The Penny Carnival ball gum was numbered and sixty of those numbers were winners along with the last ball which received the grand prize.

PENNY KING and PENNY KING TWINS The Penny King was a miniature, booth–type vendor. It was an aluminum gumball machine with a domed globe. The machine was available as a single or as the twins, which was made by joining two singles at the base. Penny King not only made machines under their own name, but also private labeled the machines for other companies. The Twin that is pictured here was made for the Flatbush Gum Co. The machines are usually marked on the bottom. There are variances in the gumball spout, where the gumball comes out. c. 1930's

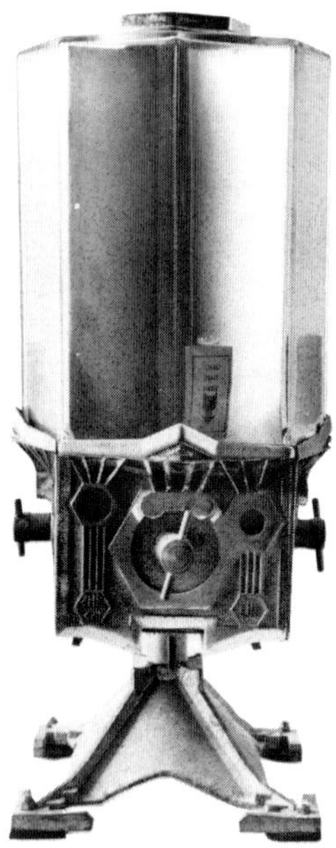

4 IN 1 An art deco stylized machine with a chromium finish that rotates on a base. The last coin remains visible for slug detection Each machine is marked "4 in 1" and "EHL." Penny King System, owned by Earl H. Lohmuller, was the first manufacturer and I believe the "EHL" represented Mr. Lohmuller's initials. In 1937 the 4 in 1 Mfg. Co. started making the machines. **Penny King System** c. 1935

PENNY PRODUCTS Each of the items in this 5–column vendor sold for a penny apiece. The machine vends mints, peanuts, gum, matches, and chocolates. It is constructed out of sheet metal and has a beveled mirror in the center. **Penny Products Vending Co.** c. 1920

PEPSIN MINT GUM The machine, which looks like the Pepsin Gum, is rumored to have been made by the Columbus Vending Company. It is similar to the Columbus Model G and L except this model has a beveled mirror instead of an embossed face. c. 1910

PEPSIN GUM Many people believe that this machine was made by Columbus Vending Co. It has a beveled window or mirror where the Columbus Model G or L has a face. Note it is even different than the Pepsin Mint Gum machine. c. 1910

PERFECTION The Perfection is exactly the same as the Chic–Mint gum machine. It has a sheet metal body and a cast iron mechanism. This is an early gumball machine. c. 1915

PERK UP and PERK UP DOUBLE This is a nickel breath pellet tray–vendor that is made out of cast aluminum. Many were also found that sell Perk Up caffeine "stay awake" pellets. I believe the wrong globes are on the double and they should look like the one that is on the single model. c. 1950's

SAMOA CHEWING GUM BOX This primitive little machine is cheaply constructed out of wood and tin. It is believed that the machine was free with the purchase of a case of Samoa Chewing Gum. This is an early machine. **Philadelphia Chewing Gum Mfg. Co.** c. 1900

PICK–A–NINNY Nickel packages of gum were dispensed by this wall–mounted vendor. It would not receive a nickel after the last package was sold. Goudey also made Big Boy Gum. I've seen a Wrigley's version of the machine. **Goudey**

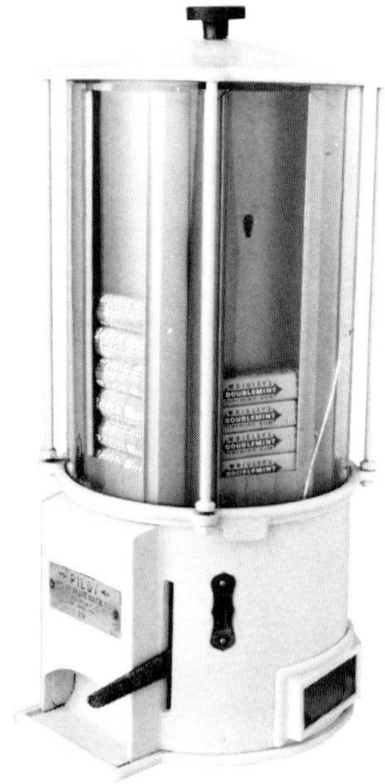

PILOT This cast iron machine vends 5¢ packages of gum or rolls of mints. The customer turns the revolving 5–column magazine to his choice. The rods that hold down the globe are on the outside. A glass window on the side shows the money in the coin box. **Pilot Machine Co.** c. 1920's

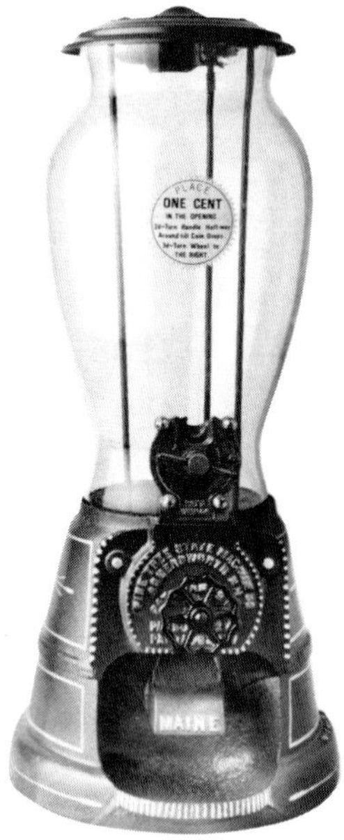

MAINE I believe that this machine was made for the Pine Tree State Machine Co. by the Advance Co. It is the same in every way as the Advance Climax except it has "Pine Tree State Mach. Co., Somersworth, N.H." embossed around the handle and the word "Maine" embossed on the chute. The following patent dates are embossed on the machine: 01–19–09, 7–2–07; 8–20–07; and 8–27–07. All of these patent dates belong to the Advance Co. **Pine Tree State Machine Co.** pat. Jan. 19, 1909

CIGARES EN CHOCOLAT This French machine vends chocolate cigars. It looks like a soldier standing at the door to a wooden guard house. He turns 90 to the right and a chocolate cigar drops into his hands, then he rotates back. He looks like a soldier at attention, however, he is holding a chocolate cigar instead of a rifle. There is another version that dispenses real cigars. **POM Distributeur**

MADAM ZITA Madam Zita may have earned her reputation as a fortune teller, but she was originally patented as a vending instrument. The patent reads "the automaton, which during the operation of the instrument, withdraws the package of chocolate, candy, or other vendible article from the receptacle wherein a number of such are held." The tension in her hands was adjustable so she could grab a chocolate bar as easily as your fortune. **Roover Bros. Co.** pat. Feb. 4, 1896

UNCLE SAM You put a penny in the coin slot in Uncle Sam's hat, then you grab the brass cigar in his mouth and crank it across his face! The hot peanuts come out underneath his beard. The machine stands 63" tall. The rectangular base and Uncle Sam's face are made out of cast iron. The machine was manufactured for the Western Sales Co. A great piece of folk art. **Mottville Mach. Works** pat. Dec. 27, 1904

NIAGRA This 4– column Niagra has a clockwork mechanism. The decal reads "Niagra Pepsin Gum and Tobler's Swiss Milk Chocolate. Turn wheel at top from you until chocolate or flavor of gum desired faces you. Then drop penny in slot and machine will do the rest. Your fortune and a piece of gum one cent." One of the greatest decals I have ever seen. **Niagra Pepsin Gum Co.** pat. Aug. 23,1904

NIAGRA According to the patent, the standing figure was to take off his hat and at the same time strike the head of the comical character at the moment when the gum is expelled from his mouth. It is operated by a clockwork mechanism. On the inside, the machine is numbered and signed Otto Schmidt, however, the patent was issued to A. Q. Allis. The only negative thing I can say about this machine is that it is not in my collection, much to my chagrin. **Niagra Pepsin Gum Co.** pat. Aug. 23, 1904

LUCAT "THE LUCKY CAT" Lucat is a cast iron cat laying on a wooden box. The tag on his forehead reads "Lucat The Lucky Cat". When you pull the handle which is an extension of the cat's tail, a numbered ticket comes out of the cat's mouth. The customer would compare his number to an award card which showed the winning numbers. Pushing a button on the side of the cat dispenses a gumball between the cat's two front feet. **Lu-Kat Novelty Co.** c. 1920

FREEPORT GUM VENDOR Because of the Goo Goo Gum ads on the side, this machine was also known as the Goo Goo Gum machine. There were two versions of the front casting available. This one has the long front window and the protruding coin entry. **Freeport Novelty Co.** c. 1899

GROVES YELLOW KID The Yellow kid, the first cartoon character, and Buster Brown were created by Richard Outcault. Captions were always written on the Yellow Kid's shirt **Grove's Yellow Kid Pepsin Gum** pat. Jul. 25, 1899

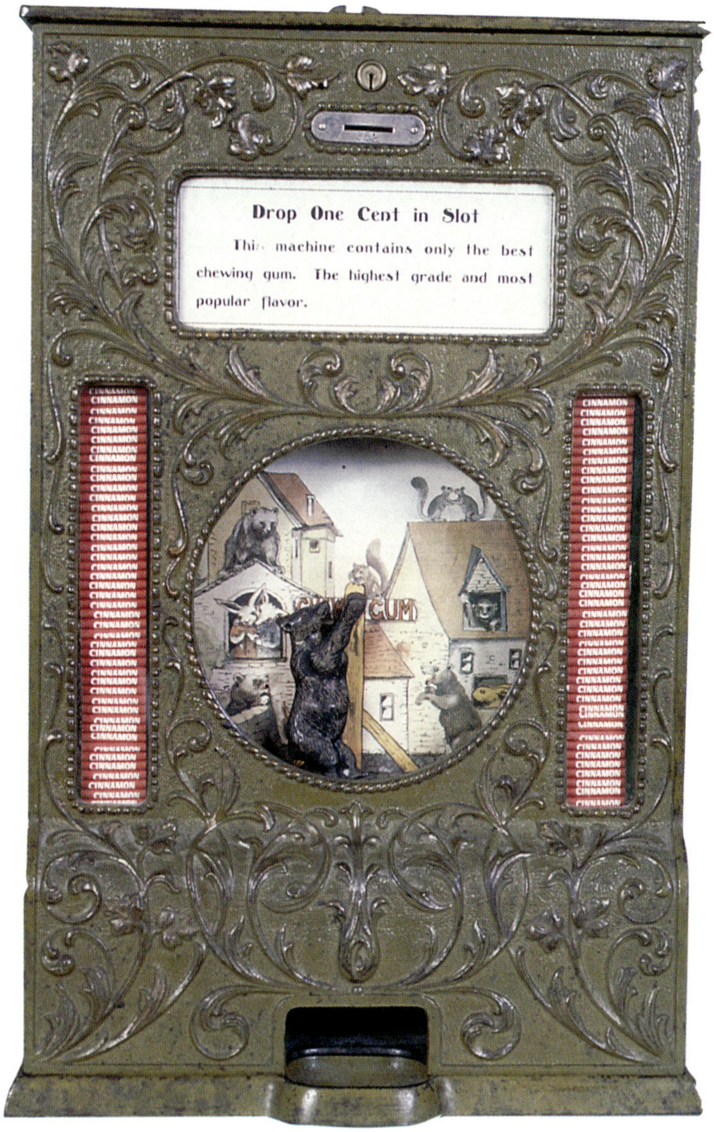

CAILLE SUNBURST Caille has the reputation of making very ornate machines and the Sunburst is no exception. Note the cherubs on the feet. The globe is etched with a Sunburst. Caille's ads state that this is the only machine on the market with an adjustable cradle that regulates the amount of peanuts or candy to be dispensed. Their ads also state "the lightest and most handsome machine of the kind on the market". As the machine weighed 34 lbs. in the box, you can imagine how heavy the competition was. **Caille Bros. Co.** c. 1909

TEDDY BEAR All the customer has to do is drop a penny in the machine. Teddy will make a complete somersault, while the machine dispenses a stick of gum. The Teddy Bear is operated by a clockwork mechanism. Not only is the machine animated, but it is one of the most ornate vending machines ever, with heavy art nouveau embossing—typical of Caille machines. **Caille Bros. Co.** c. 1909

CASE PEPSIN GUM It is easy to see how the machine was nicknamed the Fortune Teller, as the front of the machine says "Pepsin gum and your fortune, one cent." After depositing a coin, the disk in front of the fortune teller would spin, later stopping on a specific letter on the wheel that corresponds with a letter at the top of the machine. The customer would find his fortune printed after the letter. For example, the letter "O" reads "A surprise is in store for you." Only the wheel moves, the fortune teller remains stationery. **Case Chicle Gum Co.** c. 1910

MERCHANT The Gable Merchant is a beautiful example of a heavily embossed art nouveau machine. The front panel of the machine is embossed "The Merchant." This peanut vendor could operate on either a penny or a nickel. Gabel's ads stated that the more simple an automatic vending machine was, the closer it came to perfection. Gable bragged that the Merchant had only three working parts. This cast iron beauty weighed in at a mere 45 pounds. **Gable Co.** pat. Sept. 3, 1901

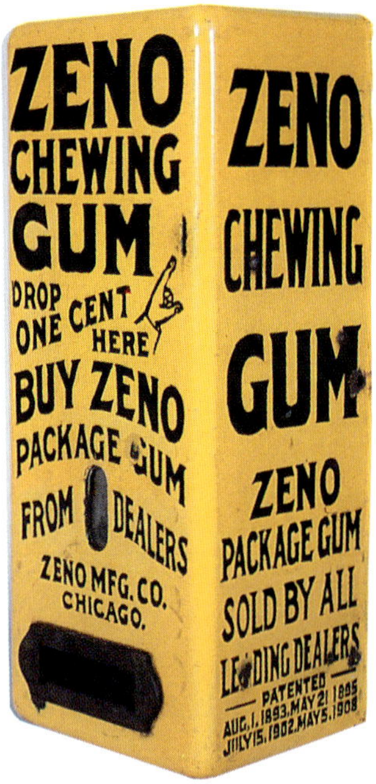

ZENO GUM These porcelain machines have black lettering over a yellow or white background,with white being scarcer. Like the older wooden Zeno, it has a clockwork mechanism. **Zeno Mfg. Co.** pat. May 5, 1908

STOLLWERCK AUTOMAT The Automat is an early 3–column chocolate vendor. The embossed front casting is made out of cast iron. It has sheet metal sides with advertising on them. **Stollwerck** c. 1890

WI-CO The machine is marked "Wi-Co." Not much is known about the machine except that it is a very early cast iron gumball machine, even the mechanism is made out of cast iron. c. 1910

HILO ROLL GUM The Hilo dispensed assorted flavors of rolled gum. The machine and ten boxes of Hilo Rolled Gum originally sold for $7.50. Hilo also made a 2–column model. **Hilo Gum Co.** c. 1905

SINGER'S GUM Note the similarity between the Singer model and the version made by Purdy-Moore. The birds are animated by a clockwork mechanism, and a bellows enables them to sing. **Singer Products Co.** c. 1928

LION The Lion was made by the Kelley Mfg. Co. as were the Kemaco and the Pix. Compare with the machines in the black and white section. **Kelley Mfg. Co.** pat. Aug. 23, 1904.

BLINKEY The original patent for a Blinkey–type machine was issued to W. J. Paul of Pittsburgh, Pa. One type of Blinkey has "Columbus Model G" embossed on it, advertising Adam's Tutti - Frutti Gum, and patented in 1907. Columbus, however, was not yet in business. Some Blinkeys had Hance stickers on them, but John Hance of Hance Foundry said that they did not make the machine. The original patent shows the Blinkey mounted on a penny drop. Since the patent was issued in Pittsburgh, where Standard is located, and Standard is the only manufacturer that I found that mounted machines on a penny drop, it is my conclusion that Standard manufactured the Blinkey. Embossed on the machine is "A Wink, A Smile With Every Gum." **Standard Gum Machine Works** pat. Nov 26, 1907

TUTTI–FRUTTI I'm guessing that Standard made this machine, too, as it has the same patent date. The letter "B" is located where others read "Blinkey." The eyes and mouth also move on this model. **Standard Gum Machine Works** pat. 1907

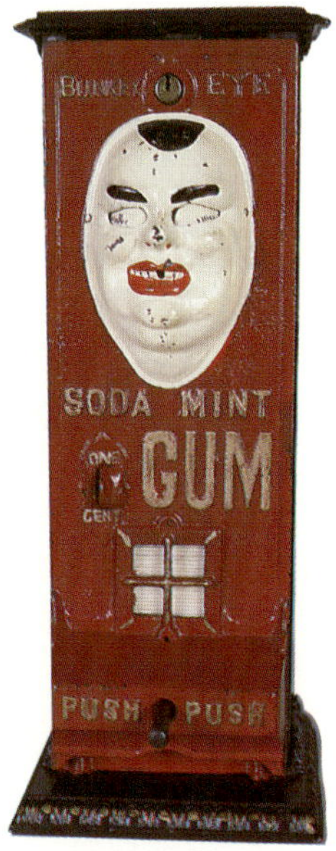

BLINKEY EYE The Soda Mint Gum Co. and Standard Gum Machine Works are both located in Pittsburg, leading me to believe that Standard is the Blinkey's manufacturer. The eyes and mouth move each time a piece of gum was dispensed. **Standard Gum Machine Works.** pat. 1907

A WINK A SMILE This is how the single version looks without the attached Daisy match machine and the penny drop. The clown's eyes move up and down while his mouth turns into a smile. The original background color was orange on this model. **Standard Gum Machine Works** pat. 1907

MODEL L "Columbus Model L" is embossed on this machine. It has a different face than the Blinkeys and the eyes and mouth do not move. I have seen these machines with Hance Foundry or Lehman & Son Co., Inc. decals on them, but they were distributors. **Columbus Vending Co.** c. 1910

LOOP THE LOOP Diavalo "loops the loop" as the machine dispenses a stick of gum. The driving force behind Diavalo is a clockwork mechanism. The machine was distributed by Caille Bros. and manufactured by the Mercer Mfg. Co. **Mercer Mfg. Co.** pat. June 28, 1904

FREEPORT DRAGONS The copper plated cast iron front is highly embossed with dragons. The case is made out of wood. I have seen two different versions. The first has round windows on the sides. The second has ornate castings on the sides where the other has windows. **Freeport Novelty Co.** 1905

FLEER'S PEPSIN GUM Fleer's used this machine that vended one stick of gum for one cent as an inducement to buy the nickel package of gum pictured on the front of the machine. Fleers invented Chiclets. In 1914 Fleer sold out to American Chicle. In 1906 Fleer was the first to invent bubble gum, known as Blibber-Blubber. Today Fleer is known for their Double Bubble Gum. **Frank H. Fleer & Co.** c. 1898

SODA MINT GUM The sign says "Chew Soda Mint Gum For Your Stomach's Sake." Early gum was advertised to correct an upset stomach. This machine has a wooden case with a tin front. It makes great country store piece. **Soda Mint Gum Co.** c. 1895

MODEL GUM SHOP Could this patent pending machine be pre Baker Boy? It has a cast iron front and roof. Note how the gumball is pushed out of the scoop by a rod. **Model Gum Shop** c. 1920's

CLOVENA It was a good thing that breath pellets had a return of 500% because the Clovena had such a small globe. **Clovena Mach. Co.** pat. Sept. 1, 1908

MAGUIRE'S GUM The machine was patented to Felix Q. Rast on Dec. 27, 1921. It is made of pot metal and is platted in copper. A beauty! pat. Dec. 27, 1921.

IN THE BAG MACHINE This machine automatically placed peanuts or candy in a bag. The nuts or candy were then eaten out of the bag as one would drink water out of a cup. This way you would not soil your hands or spread disease germs. One pound of peanuts required fifty bags. **In the Bag Co.** 1934

BAKER BOY Different automatons were available for this machine: Baker Boy, Black Man, Bell Hop, and Indian. See the black and white section. The automaton turns, scoops a gumball out of the oven, nods, and drops the gumball down the chute. **Maniken Vendor** pat. Aug. 27, 1929

STA-HOT This polished aluminum, hot nut machine has been mistakenly referred to as the Number 7 in the past. It was listed on Viking Specialty's ad page as the #7 product and the name has stuck. **Standard Appliances, Inc.** 1932

ST. NICHOLAS Even if Santa wasn't so collectible, this has to be on everybody's hit list. An early 4–column machine that dispensed St. Nicholas Pepsin Gum and Huyler's Chocolate. Note how they advertised St. Nicholas gum as an effective remedy for heartburn and indigestion. 1899

ST. NICHOLAS PEPSIN GUM The smaller St. Nicholas machine has the same wonderful Santa on the front as its bigger brother. It is a 2–column vendor, one for St. Nicholas Pepsin Gum and the other for Huyler's Chocolate. The machine has a porcelain front and wood sides. **Automatic Vending Co.** c. 1899

TROPICO CHEWING GUM This machine utilized the same mechanism as the glass domed version, however, the hardier metal casing enabled the machine to be used as a wall mount. Tropico wrappers were redeemable for premiums. **National Vending Machine Co.** pat. Feb. 19, 1907

TAFFY TOLU This is the same machine as the Chiclets or the Wilbur's Chocolate, however, this one dispensed sticks of Colgan's Taffy Tolu. With any location at all, this one column vender would have needed a lot of refilling. **National Vending Machine Co.** pat. Nov. 1, 1904

TAFFY TOLU This machine has a novel feature to induce the customer to put money in the machine. A clockwork mechanism dispenses the Taffy Tolu gum from alternate columns and the fifth penny is rewarded with gum from both columns. **National Vending Machine Co.** pat. June 24, 1902

WILBUR'S CHOCOLATE This glass–domed vendor only had a single column but was versatile. This one dispensed chocolate bars, however, others dispensed tabs of gum or chiclets. The machine was very light–weight and could easily be knocked off the counter. **National Vending Machine Co.** pat. 1904

CHICLETS This is the same machine as the Wilbur's Chocolate or the Colgan's Taffy Tolu, however, this one dispensed boxes of Chiclets. With its fragile globe and lightweight base, it is not surprising that there are so few survivors. **National Vending Machine Co.** pat. 1907

HONEY BREATH BALLS This has to be the simplest mechanism of all time—there are no moving parts! The penny goes down a track, then left. The top of the penny knocks the breath ball off the chute and out, thus dispensing four balls each time. See gumball version. **Gravity Vending Mach. Co.** pat. Dec. 1, 1903

HAPPY JAP When the customer deposits his penny, the Happy Jap automatically pushes the gum through his teeth. Note the gum slots on each side of his mouth. He is constructed of cast iron and has a clockwork mechanism **White Vending Machine Co.** pat. 1902

ZENO GUM This machine is the same as the wooden Zeno with the clockwork mechanism, however, the wood is covered in tin with advertising on it. **Zeno Mfg. Co.** pat. Aug. 1, 1893

BERGER The front and top of this early bulk vendor are made out of cast iron, while the rest of the body is painted sheet metal. The legs are actually faces. **Paul E. Berger Mfg. Co.** pat. June 21, 1904

SMILIN' SAM How many of you have ever seen this variation of Smilin' Sam? Even though he is wearing a top hat full of peanuts in a shell, he still dispenses shelled peanuts through his tongue like the other models. **General Mdse Co.** 1931

SMILIN' SAM You have to pull Smilin' Sam's tongue out in order to get your ration of peanuts. The raised coin entry is correct. Note the similarity between this aluminum Smilin' Sam and the much earlier cast iron Happy Jap. **General Mdse. Co.** 1931

SMILIN' SAM Smilin' Sam from Alabam' The Salted Peanut Man, was available as a white man, black man, Indian, Chinaman, or polished aluminum. Where did the variations all go? Flat coin entry is correct. This machine has been reproduced several times. **General Mdse. Co.** 1931

REPETTIS / FLEERS The only mirror on the subway platform was on the machine thus, attracting many potential customers. **Automatic Vending Co.** pat. June 27, 1899

BROUSE'S PEERLESS Another early 4–column vendor, which advertises Brouse's Peerless Fruit Chewing Gum. **Brouse's Peerless Fruit Chewing Gum.** c. 1890

CHICLETS The porcelain sides were utilized to advertise whatever was in this machine; such as Chiclets, Dentyne, and Stollwerk's Chocolate. **Autosales Gum & Chocolate Co.** c. 1899

FROZEN MINTS By changing the side panels, Autosales was able to promote whatever was in the machine. **Autosales Gum & Chocolate Co.** c 1899.

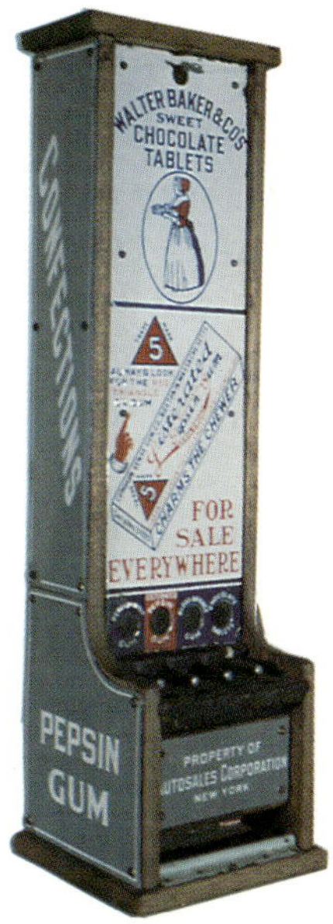

ADAMS' Adams is known as the man who first popularized vending machines in the U S. He started putting his machines on the N.Y.C. raised subway platforms in 1888. **United States Slot Machine Co.** pat. Dec. 4, 1894

WILBUR'S CHOCOLATE The porcelain sides picture the Wilbur's trademark of a cupid with a cup of cocoa. The 4–column vendor dispensed chocolate and two flavors of gum. **Champion** pat. May 16, 1899

WALTER BAKER CO. This 4–column vendor is marked Autosales Corp. which probably came after the Autosales Gum & Chocolate Co. **Autosales Corp.** c. 1910

STOLLWERCK These are some of the earliest machines. They were patented by Stollwerck, a German, and assigned to John Volkman, a U. S. citizen. Wood grained porcelain makes this an unusual example. **Volkmann Stollwerck** pat. 1888

PULVER KOLA-PEPSIN GUM (left) **AND PULVER CHOCOLATE COCOA AND GUM** (right) and side view (center) The company name, the Pulver Chocolate and Chicle Mfg. Co., definitely sounds much more appetizing than their previous name, the Pulver Chemical Co. Both machines have a beautiful, highly embossed lithographed tin casing. The side panels on both are the same, however, the front panels are different. The upper front panel on the Kola-Pepsin shows a picture of a 5¢ package of gum which was only available over the counter. The machine was a 1¢ machine and did not dispense the 5¢ package of gum. On the bottom panel of the Kola-Pepsin, the pointing hand is level with the coin entries. The hand is lower and pointing up to the coin slots on the Chocolate Cocoa machine. Also, pictured is a package of Pulver Cocoa with a little girl on it. The characters that are found in these models are: The Yellow Kid, Buster Brown, Foxy Grandpa (aka The Professor) and Uncle Sam. **Pulver Chocolate & Chicle Mfg.** pat. May 30, 1899

KOLA PEPSIN GUM Even though this machine has the same patent dates as the other lithographed tin machines it was made be the Pulver Chemical Co., which sounds disgusting. This machine is not embossed like the other two machines. All the panels on this machine are different than the other two machines. The side panels on this vendor are absolutely fantastic. They show two adorable children and a box of 5¢ packages of Pulver Kola–Pepsin Chewing Gum. **Pulver Chemical Co.** pat. May 30, 1899

WRIGLEY'S This machine had instant recognition, as everybody knew the Wrigley's man because Wrigley was such a prolific advertiser. The european machine vended Wrigley's P K Gum c. 1915

PULVER TOO CHOOS There are two different tall porcelain–cased Pulvers. The puppet inside was dressed in fabric clothing. Characters would be the Yellow Kid, Buster Brown, Foxy Grandpa (aka The professor), or Uncle Sam. **Pulver**

PULVER KOLA–PEPSIN This is the second of the tall porcelain cases. Note that it has completely different porcelain panels than the Pulver Too Choos. It is scarcer than the Too Choos. **Pulver** c. 1910

BEECH-NUT GUM The Beech-Nut gum machine is a porcelain wall machine. It is missing the knob. They used the same little girl in their early advertising campaigns. c. 1920's

DR. KING'S FRUITLETS (left) **AND RED CROSS** (right) Note that the Fruitlets has Dr. King's on the side, the same as the Red Cross. Both machines are Canadian. They are believed to be made by Pulver for the following reasons: the machines are exactly the same size as Pulvers, they use the same puppets, all of the parts are interchangeable with Pulvers, and Pulver's proximity to Canada. c. 1910

MILLS AUTOMATIC The Autosales Corp. has the same machine with different advertising in the porcelain and a mirror instead of the window that is on the Mill's. This is 4–column gum and chocolate vendor. **Mill's**

TRUCKY The truck drives in a circle. As the truck starts, the man on the left turns, dispensing gumballs onto a conveyor belt that unloads into the hopper. Then the truck trips the hopper, which drops the gumballs into the truck. Finishing its path, the truck drops the gumballs into a hole that dispenses them to the customer. The older version had a metal truck and operated on a penny, while the newer version has a plastic truck and gives several gumballs for a nickel. **A M Co.**

SCOOPY GUM VENDOR This Scoopy has three vending compartments but only the middle one operates Scoopy. Scoopy is a cheaper imitation of Manikin Vendor's Baker Boy. Note how the automaton is dressed. **A M Co.**

SCOOPY GUM VENDOR This is a cheaper copy of the Maniken Vendor (Baker Boy). I have seen it with "A M Co." on the chute cover and without. I have also seen this version advertised as being made by the Harris Gaylord Mfg. Co. **A M Co.** c 1920's

SCOOPY I have seen this Scoopy Gum Vendor with a couple of different characters. The body is made of sheet metal. If it has A M Co. on the chute cover the advertisements say they were made by the Scoopy Mfg. Co. **A M Co.** c. 1920's

A M CO. Note how this character is different than the Scoopy machines. Instead of getting the gumball out of the oven like a baker, he pulls the gumball from over his head. It also has a different character. **A M Co.** c. 1920's

WOODEN PULVER This is the larger of the two wooden pulvers. The early wooden pulvers are 1–column vendors. The only automaton that is correct for this machine is the Yellow Kid, who was the first comic strip character. **Pulver Chemical Co.** pat. May 30, 1899

WOODEN PULVER This is the smaller, skinnier version of the wooden Pulver. Once again, the only correct animated character is the Yellow Kid. This 1–column vendor dispensed 1¢ tab gum, some of which were wrapped in wrappers that were redeemable for a 5¢ package. The sign reads, "If you're the feller as what was lookin' for a chew, put a penny in the slot and I'll give it ter you. The Kid." **Pulver Chemical Co.** pat. Dec. 29, 1897

KOLA PEPSIN GUM This is the tall, painted Pulver case. It has black, embossed printing over an orange background. It, too, is a 1–column vendor. The correct automatons are the Foxy Grandpa (aka the Professor), the Yellow Kid, Buster Brown, Happy Hooligan, and Uncle Sam. The Foxy Grandpa is pictured. **Pulver Chemical Co.** May 30, 1899

KOLA–PEPSIN GUM This is a very unusual Pulver for several reasons. First of all, the case has diamond shaped windows on the side. Secondly, the character inside is Happy Hooligan, who is not seen very often. Note how the Happy Hooligan pulls down the lever that dispenses the gum. This early Pulver is a 1–column vendor. **Pulver Chemical Co.** pat. May 30, 1899

PULVER The front of both cases is the same, however, one is advertising Pulver Too Choos Gum on the side. I believe that model is older. The cases were available in the following colors: red; blue; green; yellow; white; and orange. The characters found in the short porcelain cases are: The Yellow Kid, both in fabric clothing and die–cut, formed tin; cop and robber; cop directing traffic; clown and Woody Woodpecker. **Pulver Mfg. Co.**

PULVER This painted case is the last one that Pulver made that housed an animated character. The yellow crinkle finish is more common than the red. The plastic Woody Woodpecker and the clown were the characters used in this machine. **Pulver Mfg. Co.**

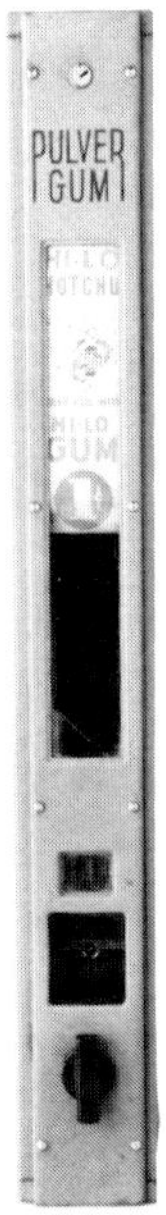

PULVER The first model shows three tubes of liquid. Each time the customer gets his gum it causes the liquid to rise in the tubes. The object is to see how high he could make the liquid rise. The second model has a mirror instead of tubes. Both have red porcelain cases but I have heard of a yellow painted model. **Pulver Mfg. Co.** c. 1950

PULVER HOTCHU Both machines are a 1–column vendor dispensing HotChu Gum. One model says Hi–Lo and the other says Pulver's HotChu. The first machine is finshed in an orange and green crinkle–paint finish while the second is finished in chrome. **Pulver Mfg. Co.**

YELLOW KID (left) **BUSTER BROWN** (right) The Yellow Kid was the original comic strip character and first appeared in the *N. Y. World* on July 7, 1895. He was created by Richard F. Outcault who also created Buster Brown in 1902. Both were dressed in fabric clothing and both were found in the tall cases, however, the Yellow Kid was also found in the short porcelain case. **Pulver**

FOXY GRANDPA This comic character, done by C. E. Shultze, pseudonym Bunny, debuted in the *N. Y Herald* on Jan. 7, 1900. Incorrectly called the professor by many collectors.He wore fabric clothing.**Pulver**

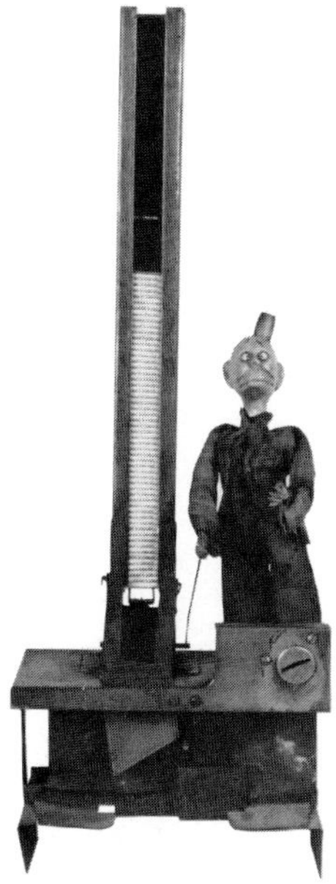

HAPPY HOOLIGAN Happy Hooligan, the Irish tramp with the tin can hat,debuted in the comic strips on Mar. 26, 1900. He was created by Fred Opper. He wore fabric clothing and was found in the tall cases. **Pulver**

UNCLE SAM Uncle Sam was found in the tall cases. Be careful when buying one of these, as some of them are not original. He wore fabric clothing. **Pulver.**

COP AND ROBBER The cop hits the robber over the head once with a billy club, the robber's eyes go up and down, and he sticks out his tongue three times. The figures are made of a composition material. **Pulver**

COP DIRECTING TRAFFIC The policeman is holding a stop and go sign. He is made of a composition material. He turns about 45° to the right and then back again. He was found in the short porcelain cases. **Pulver**

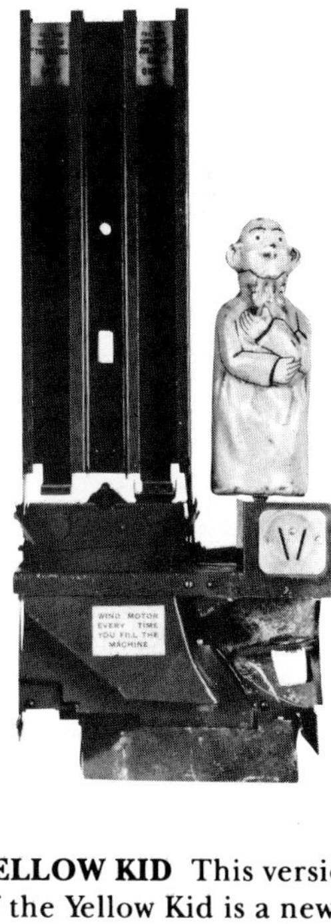

YELLOW KID This version of the Yellow Kid is a newer one and was made for the shorter porcelain cases. He is a die-cut, molded, lithographed tin character. **Pulver.**

CLOWN The clown was one of the last Pulver figures. He was made out of a composition material. He turns about 45° to the right and then back again. He was found in the short porcelain or painted cases. **Pulver**

WOODY WOODPECKER Woody was a cartoon character created by Walter Lantz in 1941. Woody was one of Pulver's last characters. The red plastic character was found in the short porcelain or painted cases. **Pulver**

BIRDIES The sign says "Drop 1 cent, hear and see Birdies sing – perform and give you pure – sanitary delicious Gum." A clockwork mechanism animates the birds while a bellows enables them to sing. **Purdy–Moore Mfg. Co.** c. 1928

ADAMS' PEPSIN TUTTI–FRUTTI GUM Thomas Adams is generally considered as the man who popularized vending machines by his extensive use of them. This is another vendor used by Adams to market his Tutti–Frutti Gum. **Railway Automatic Sales Co.** c. 1898

SILVER COMET This is a 1–column stick gum vendor that is made out of sheet metal. It utilizes a coin slide and the gum comes out the side. **Redco Products Corp.** c. 1925

REGAL This cast iron machine, the first Regal, was finished in crackle–paint. It has also been found with an Ambassador tag on the flap. **Regal Products Co.** c. 1930's

REGAL A cylinder globe and a large lid was used on this aluminum version of the Regal.. The base is the same as the model with the tear–shaped globe. **Regal Products Co.** c. 1940's

BLUE REGAL The base of the Blue Regal is the same as the model with the cylindrical globe. If you have the tear–shaped globe you need the small lid. **Regal Products Co.** c. 1940's

CASH TRAY Their ad reads, "sets right on the bar and slides from one customer to the next, resulting in sales of 600% to 800% greater than a stationery machine." In 1946 it became the Adams–Fairfax Corp. **Reliable Nut Co.** 1934

RING DING BOMB The all plastic Ring Ding Bomb was a 5¢ capsule machine. It dispensed bubble gum and a toy for a nickel. The machine was free with the purchase of the capsules. **Ring Ding** c. 1950's

RING DING With a large order of gum, the customer got the machine for free. This machine was meant for store use as it was not rugged enough for use by an operator. **Ring Ding** c. 1950's

SPACE SHIP The Space Ship was another "freebie" with a large order of gum. The all plastic vendor must have dispensed small pieces of gum because the customer got 6 pieces for 1¢! **Ring Ding** c. 1950's

2 COLUMN VENDOR D. Robbins Co. was a large N.Y.C. distributor of coin operated machines. This was one of their own machines. A 2 column gum vendor that holds 200 pieces of stick gum. **D. Robbins Co.**

EMPIRE This machine looks very similar to a Masters but was made by the D. Robbins Co., New York's largest vending machine distributor. It was available as a bulk vendor or a gumball, profit–sharing or hot nut machine. The Empire State Building is embossed on the front. **D. Robbins Co.** 1932

PEACHES AND CREAM The Peaches and Cream is a 2–column vendor with a clockwork mechanism. It dispensed tab gum. **Robertson Sales Co.** 1906

MECHANICAL SALESMAN The patent number is the same as for the Maniken Baker Boy, but this is a bulk vendor rather than a gumball machine. When a coin is deposited, the puppet nods his head, fills his basket, and delivers the merchandise with a nod of thanks. **Ro–Bo Sales Corp.** pat. Aug. 27, 1929

PANSY GUM Pansy gum seemed to have always been sold with a fortune and a love letter. The three Pansy gum vendors are quite different, but are all examples of early vendors. This version has a stenciled, cast iron front and a wooden case. **Roth's Pansy Gum** c. 1905

ROYAL VENDOR The Bicycle Bell and Thumb Press were nicknames used for this machine before its manufacturer was known. It only came in a gumball version and had a metal body with a chrome plated, brass mechanism that works similarly to the Advance gumball mechanism. **Royal Gum Co.** pat. Nov. 2, 1915

CHEWING GUM A mirror is on the front of this sheet metal 1–column vendor. It could dispense a variety of products but this one dispenses gum. The window in the front allows the customer to make sure there is merchandise in the machine before inserting his coin. **Ryede Specialty Works** c. 1920's

WILBUR'S The porcelain signs on the sides advertise Wilbur's Sweet Clover Chocolate. The 4–column machine vends gum or chocolate. The left half of the machine is impressed "Model G Pat. April 16, 05; Feb 22, 1910; May 21, 1912; July 30, 1912" and the right half "Pat. Pending." **Ryede Specialty Works**

STOLLWERCK SAMPLER A clockwork mechanism rotates a vertical wire corkscrew that drops a 5¢ chocolate bar. The case is made of oak. A second variation of this machine allows the customer to see the corkscrew and chocolate. **Schilling, Stollwerck & Co.** c. 1905

FOUR WAY This cast iron machine looks more like a watermelon than a vending machine. The machine had four compartments, divided by glass partitions with four separate coin mechanisms and coin entries. **Selective Vending Machine Co.** c. 1934

SELECT-O-VEND This multi–columned vendor dispensed tab gum or chocolate. The customer turns the rotating magazine to the selection of his choice, inserts his penny and pulls the knob. c. 1945

SHIPMAN This machine has a single lock for the top and the bottom. The screw type lock in the bottom screws to a rod attached to the cap. This porcelain machine was available with or without a hot nut flasher. Note pull chain instead of a handle. **Shipman Mfg. Co.** 1932

SILVER KING Three different manufacturers made silver Kings: Automat Games from 1935 to 1939, Automatic Games from 1939 to 1945 and Silver King Corp. after 1945. The aluminum Silver King with the curved chute with "Try Some" was introduced in 1946. **Silver King Corp.** 1946

HOT NUT A ruby hobnail glass top flashes atop this machine. The rest of the machine was the same as the regular aluminum machine. It was available with or without the side cup dispenser. **Silver King Corp.** 1947

MUSICAL BALLERINA For a nickel the customer not only got a handful of nuts but got to see the ballerina dance to music. **Silver King Corp.** c. 1950

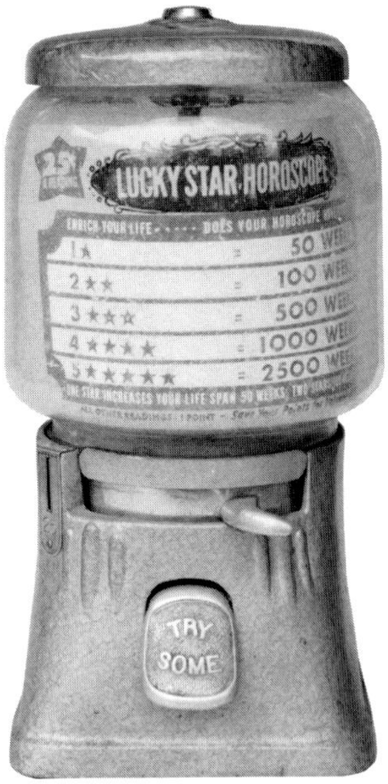

GIANT ACE This is a conversion from the standard 5 lb. tapered globe vendors to a 7 lb. globe with an extra large top. **Silver King Corp.** 1953

TAB GUM Silver King's tab gum machine held a total of 200 pieces of gum or chiclets in its four columns. It had the company's patented coin shooter which prevented the insertion of foreign objects that would put the machine out of order. **Silver King Corp.** c. 1954

SILVER QUEEN This 5–column machine dispensed chocolate or tab gum. Silver King had a 4–column bulk vendor called the Silver Queen and this machine is probably one of theirs as well. This model has also been seen with the name Lawrence Mfg. on the machine.

MODEL A There was a proliferation of new machines in the mid 30's due to the repeal of prohibition. This one was usually finished in red, green, or yellow porcelain but was also available in chrome. The geometrically embossed 8–sided globe gives this machine an art deco look. **Simmons Coin Machine Co.** 1934

SIMPSON I believe that this is the machine that the September 15, 1908 patent date, found on Columbus and Ad–Lee machines, is based on. Check under Columbus for further explanation. **R. D. Simpson Co.** c. 1907

LINCOLN There is no way that this machine is the original Simpson and forerunner to the Columbus Model A as many think. I believe that honor goes to The Simpson. It had a formed steel or Columbus–type lid. The Self–Serv Mfg. Co. advertised the machine as the Lincoln. **R. D. Simpson Co.**

SIMPSON JR. This cast iron, miniature vendor dispensed breath pellets or peanuts. It has been seen with an oval or a round globe. The lid is formed metal. I've also seen this machine advertised as the Supreme Jr. and the Superior. **R. D. Simpson Co.**

MODEL A This is the big brother to the Simpson Jr. It came with either a $4\frac{1}{2}$ or 6 lb. round globe. Since it was private labeled, this machine was advertised as the Supreme or the Superior. It has a cast iron body and a formed metal lid. **R. D. Simpson Co.**

FAMOUS 1-2-3 VENDOR (aka Leebold) The turntable was interchangeable to accommodate the following combinations: 1-2-3, Pen–Nick (1 ball for a penny and 5 or 6 balls for a nickel), Double Value (2 gumballs for a penny) and Straight (1 gumball per penny). **R. D. Simpson Co.** pat. Jan. 1, 1923

"WELL, HERE WE ARE" This is a convertible vendor that could vend candy, gum, or salted nuts. It was constructed of iron and copper which was then heavily nickel plated. Two sizes of globes were available. **R. D. Simpson Co.** 1927

SUPER VENDOR Some collectors call this the "Castle Top" because of the lugs on the top of the lid and the coin entry. Only the midsection makes this machine different from the other Super Vendor. This one is lugged. There were many interchangeable parts on the different models. **R.D. Simpson Co.** c. 1929

SUPER VENDOR Cadmium plating under the crackle paint was used to rust-proof this machine. It was an adjustable vendor that could dispense ball gum, peanuts or marbles. Other versions included two balls of gum for 1¢, 5 balls for 5¢, and a 1–2–3 machine. **R. D. Simpson Co.** c. 1929

CLIMAX This machine is a bulk vendor. All of the ones I have seen have been painted. Note how the chute flap on this one is different and pulls down. Some flaps read "The Climax" and others just read "Pull." The lid is lugged and there are other interchangeable parts. **R. D. Simpson** c. 1931

THE ARISTOCRAT This bulk vendor could dispense items without crushing according to the ads. It was made of cast iron and brass and plated in chromium. Two sizes of globes were available. The lid, globe, rods and mechanism are interchangeable with other models. **R. D. Simpson Co.** c. 1935

DERBY CONFECTION VENDOR This was a combination confection and amusement machine available in gum or peanut versions. The horses rotate with each play; the one stopping at the judge's stand is the winner. The machines were made in cast iron or aluminum. "Derby Confection Vendor" is embossed on the lid, however, I have seen one machine embossed with "Simpson Derby Confection Vendor." The machine on the right has a different lid than most models and its marquee reads, "Derby Confection Vendor". The mechanism also works differently. The decal reads, "Race Park, No Premium for Merchandise, Amusement Only." **R. D. Simpson Co.** 1940

SIMPSON VENDOR This chrome plated bulk vendor was available in two globe sizes and many of its parts are interchangeable with other Simpsons. Note how it looks like either the Super Vendor ot the Derby Confection Vendor without the horses. **R. D. Simpson Co.** c. 1930

DOUBLE COMMANDER There is also a single and a triple commander. The name was undoubtedly influenced by the war. The globes are 5–sided and irregularly shaped. The base is also angled. The machine was also advertised as the Simpson Twin Vender. **R. D. Simpson** 1941

SODA MINT GUM Resembling a dye cabinet more than a gum machine, this early machine has a coin slot that closes when empty. I believe that the Soda Mint Gum Co. also used the Blinkey Eye Dispenser made by the Standard Gum Machine Works. **Soda Mint Gum Co.** c. 1905

SPARTAN HOT FLASH Even though the metal tag on this cast iron machine reads, "Spartan Hot Flash", there seems to be no question that the Hance Mfg. Co. of Westerville, Ohio actually manufactured this machine for them. **Sparta Mfg. Co.** c 1920's

IDEAL COLD PEANUT MACHINE Ads boasted that this machine was dust proof. I assume it was due to the spring loaded chute cover. Construction is of aluminum with an adjustable mechanism and came with either a 5 lb. or 7 lb. globe. Penny and nickel versions were available. **Specialty mfg. Co.** c. 1930

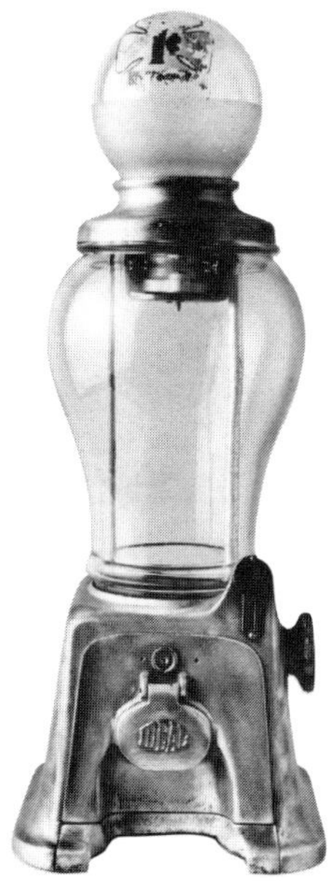

IDEAL HOT NUT MACHINE The hot nut version is the same as the Cold Nut with the exception of a heater contained in the base and a light on the top. The ads claimed that the electrical costs were 1.25¢ for 10 hours. Available with a 5 lb. or 7 lb. globe and in a penny or nickel version. **Specialty Mfg. Co.** c. 1930

PETITE VENDER This is the breath pellet model that holds 2 lbs. of merchandise. There are also gum and peanut versions of this model. All are made of cast aluminum. **Specialty Coin Machine Builders.** c. 1934

STA–HOT Note how the "S–T–A" in Sta–Hot uses the first three letters in *Standard.* This unusual hot nut machine is made of aluminum. Viking Specialty, a distributor, listed this machine in their catalog only as the seventh product on the page and the name, Number 7, has stuck. **Standard Appliances, Inc.** 1932

PEPSIN GUM A penny would buy a stick of gum and your fortune from this wooden dispenser. I believe this company also made the Blinkey–type machines. **Standard Gum Machines Works** c. 1905

LOG CABIN DUPLEX The Log Cabin Duplex is a two compartment bulk vendor. According to the September 1931 *Automatic Age,* Log Cabin penny vendors were made in Cedar Rapids by the Standard Mfg. Co. **Standard Mfg. Co.** c. 1930

HOT NUT All of the ads I have seen for the Stanley Hot Nut were by the Silent Sales Vending Co. of Philadelphia, Pa. "The Stanley" is impressed into the base. Note the similarities to the Simpson and Columbus machines. **Stanley** c. 1925

STANLEY The Columbus globe on this machine is incorrect, however, it does take a round globe. This was also a hot nut model. "The Stanley" is impressed in the base. I believe that Simpson made this machine. **Stanley** c. 1925

STAR PEPSIN CHEWING GUM This early wooden machine claimed that Star Pepsin Gum would cure indigestion, dyspepsia and heartburn. Patent dates appearing on the machine are: Oct. 18, 1887; May 17, 1889; and Aug. 23, 1892.

STEWART & McGUIRE The Empire State Building, where Stewart and McGuire's offices were located, inspired this modernistic nut and bulk vendor. The all glass panels could be replaced by metal panels with small glass openings. **Stewart & McGuire.** 1935

ADAMS GUM The hexagonal shaped machine is a 6–column vendor. The famous Adams gum was wrapped in United profit sharing coupons that were redeemable for free gifts. Stewart & McGuire billed themselves as the world's largest distributor of automatic cigarette machines. **Stewart & McGuire.** 1937

STOLLWERCK A highly embossed, cast iron casing characterizes this early 2–column chocolate vendor. A nude baby is sitting on the roof. The detail and beauty of this machine is exceptional. **Stollwerck** c. 1890

STOLLWERCK Stollwerck, a German, assigned his patents to Volkmann in the United States. The 4–column vendor has an oak case and ruby red windows advertising Stollwerck Chocolate Confections. **Volkmann, Stollwerck and Co.** Pat. Sept. 27, 1892

STOLLWERCK BROS. Stollwercks are early examples of commercial vendors. Note similarities between Stollwerck machines. This one has a mirror instead of the ruby red glass. The Frozen Mints decal was advertising one of the selections in the machine. **Volkmann, Stollwerck and Co.** Pat. Sept. 27, 1892

TWIN MERCHANDISER This bulk merchandiser was designed as a side addition to the Univendor, a candy bar machine. It has a single base. Even though the globes seem to be the same, the Stoner globe has circular ends and the Northwestern 33 Jr's globe has 8–sided ends. **Stoner Corp.** 1941

FRESH GUM This 6–column tab gum vendor is made of formed metal. **Stoner Mfg. Corp.** c. 1940's

HOT NUT This aluminum hot nut machine has a metal tag on it that reads, "Machine Owned and Leased by C. D. Stover, Waterloo, Iowa. 3–1586 For Service." It utilized a coin slide that had a patent date of May 9, 1933 on it. **C. D. Stover** c. 1933

SUN This cast iron machine looks exactly like the Chicago except it has "Sun" embossed on the sides. It has the same patent dates as the Chicago so it is believed to be made by the Coleman Hardware Co. and was patented by B. M. Davis. **Coleman Hardware Co.** Pat. Sept. 8, 1908 and Feb. 9, 1909

SURE The Sure is a King Jr. without the ashtray. Ads said, "39¢ worth of Sure Bee–Bees pays 440 pennies" – not a bad return. Sure Bee–Bees were candy pellets for the breath. The same machine came in a gumball version. **Automatic Games** 1940

MAGIC VENDOR Townsend Mfg. Co. advertised the Magic Vendor in the May '39 *Automatic Age.* Commonly known as the Al Hoff because the operator's name, "Al Hoff," is embossed on the coin door. This cast iron and sheet metal machine was advertised as having "the new turret top" with all of the mechanical parts built right in the top. It works the same way as the Calvert Vendor. The Al Hoffs are all 1–2–3 machines but some have a removable cap over the vending wheel. **Townsend Mfg. Co.**

WRIGLEY'S The little window in this 1–column vendor was used by the operator to advertise the contents of the machine. **Transportation Vendors, Inc.**

SNACKS GUM VENDOR This tab gum vendor was designed to compliment the Snacks bulk vendor. It also could be mounted on a cash register, wall or other vending machine. **Trimount Coin Machine Co.** pat. 1939

SNACKS This modernistic three compartment vendor was available in 1¢ or 5¢ versions. It had three coin entries and mechanisms and could be mounted on the counter, stand, or wall. **Trimount Coin Machine Co.** 1936

CHALLENGER The three separate coin entries on this machine were designed to be reversible, 10¢ on one side and 5¢ on the other side. It is made of brushed aluminum and has a blinking light mounted inside its top. I have seen some machines that have the name "Ajax" on the nameplate instead of "Challenger." **Tropical Trading Co.** 1947

TRUE BLUE Even though the decal states, "Manufactured by True Blue Gum Co. of Lansing, Mich", there is no doubt that the machine was actually made by the Automatic Sales Co., also of Lansing. This machine is more sophisticated than the first model. The earlier model had a push rod that made it easy for the customer to push it off of the counter, while this model has a much more intelligent handle on the side. Other than the handle, the two machines are the same. **True Blue Gum** 1906

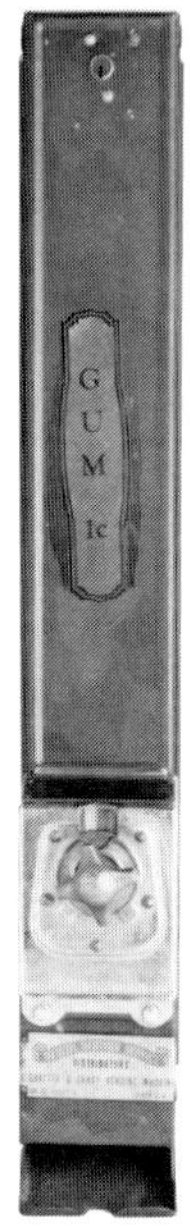

GUM This machine has an Advance mechanism on it with the following patent dates of 11–6–12 and 4–24–23. The machine was very versatile and could have just as easily been used as a prophylactic or candy machine. **Uneeda Vending Service Dist.**

MODEL A (left) **MODEL C** (center) **MODEL D** (right) The Model A and C were available as a 1–2–3, 2 for 1¢, or 1 for 1¢ while the Model D was specially built to dispense one large gumball for a penny. Note all three use the same coin mechanism. The Model A 1–2–3 had a window to show the customer how many gumballs his next penny would buy. Universal apparently private labeled the machines to Modern vending Machine Co. of Haverhill, Mass., National Gum Co., Inc. of Newark, N. J., Remington Service Machines, Inc of Ilion, N. Y., and Unxld Gum Co. of Easton, Pa. **Universal Sales Machines Co.** c. 1925

NUT KING An unusual 1¢ package gum dispenser, this machine was manufactured by Oak for Universal. The Liberty Bell is embossed on the globe and flap instead of Oak's acorn logo. **Universal Vendors**

UNIVERSAL A Liberty Bell is embossed on the flap but undoubtedly Oak Mfg. Co. is the manufacturer. Many manufacturers would private label a machine if the order was large enough. **Universal Vendors**

UNXLD This machine is believed to made by Universal Sales machine Co. for the Unxld Gum Co. It is exactly the same as Universal's 1–2– 3 except the front is embossed **Unxld Gum Co.** c. 1925

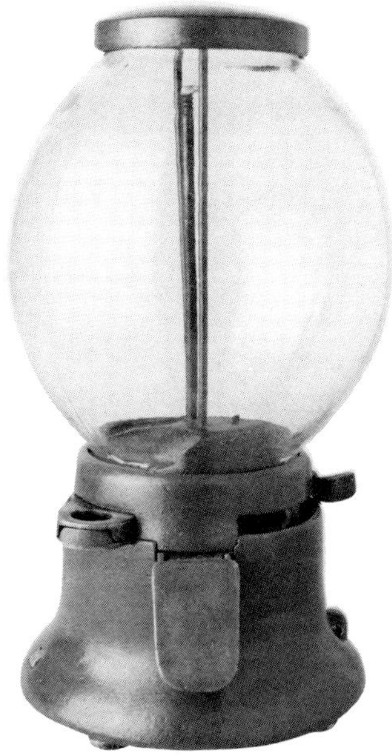

VALENTINE Embossed on the bottom of the machine is "Valentine Produce Co., Willowlake, S.D." It was likely private labeled for them by Lee Co. It has the same unusual coin mechanism as the Lee. **Lee Co.**

VELVET MINT It is a little difficult to get your fingers in the hole to retrieve your gumball. This early brass machine has a coin entry and knob on the side.

VEND–ALL Hawkeye Novelty Co. made this machine for Vend–All. The profit sharing decal reads, "When Bell Rings Free Portion Pull Lever Again". Hawkeye machines gave a free portion after every ninth operation until New York outlawed them as gambling machines. **Vend–All** c. 1931

GUMBALL The Vendex is a booth type machine. This is an earlier Vendex gumball machine, as it has the smaller round cast iron base. The gumball versions either had a painted cast iron or a polished aluminum base. It has a wire type lever and a domed globe. Most of the bases are marked "Vendex." **Vendex Co.**

PEANUT (left) **GUMBALL** (right) Notice that these machines have squared off bases for additional stability. The peanut version is made of chromium plated cast iron and the gumball machine is made of either painted cast iron or polished aluminum. The chute flap is small for a peanut vendor. The bottom of the base is usually marked "Vendex Co." **Vendex Co.** pat. 1935

VENDEX 520 This cast iron machine was advertised as the interchangeable jar ball gum vendor. The operator would bring ready filled jars and switch the full jar for the empty. **Vendex Co.** pat. 1935

GUMBALL This Vendex machine is a complete departure from anything they had done in the past. The cast iron base is much larger and heavier than before. The lever is much more substantial than the smaller booth machines. It had an interchangeable jar–type globe. **Vendex Company**

PEANUT This is an early cast iron peanut machine with a tray built into the base. It utilized a periscope coin entry and a lantern style globe. **Vending Machine Co. of America** c. 1910

BREATH PELLET This is an unusual small cast iron vendor meant to dispense breath pellets or small candies.Note the unusual vending mechanism. **Vending Machine of America** c. 1910

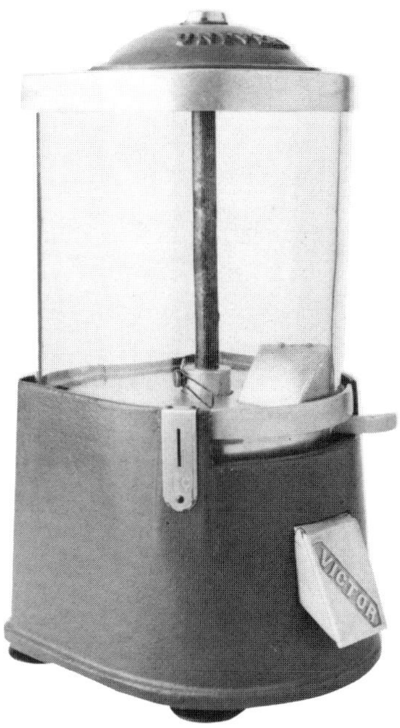

UNIVERSAL This bulk vendor seems to be Harold Schaef's first machine as Victor Vending Corp. "Universal" is embossed on the lid. I have seen some of the chute covers marked "Victor" and some are blank. **Victor Vending Corp.**

CHIEF The Chief,Victor's second machine was a single–compartment bulk vendor. Victor also made the Stop and Shop which looks like a Triple Chief. **Victor Vending Corp.** 1939

CHALLENGER This is a different bulk merchandiser that was available in porcelain or enameled cast iron. Victor changed the name of the Challenger to the Esquire in 1940. **Victor Vending Corp.** 1940

SPECIAL DELUXE The decal on this machine is wrong; it should be a Special Deluxe decal, not Esquire. This bulk vendor was Victor's top of the line machine. It was finished in porcelain and trimmed in chrome. **Victor Vending Corp.** 1939

TOPPER Topper was billed to vend everything. It came in an enamel finish or porcelain for an extra 50¢. This machine had a cylindrical globe. It was the start of Victor's most successful bulk vendor. **Victor Vending Corp.** 1938

MODEL K This machine is sometimes referred to as the sidewinder. The ads claim that the machine was made of porcelain but I have seen a painted one. Note how the handle is on the side. **Victor Vending Corp.** 1947

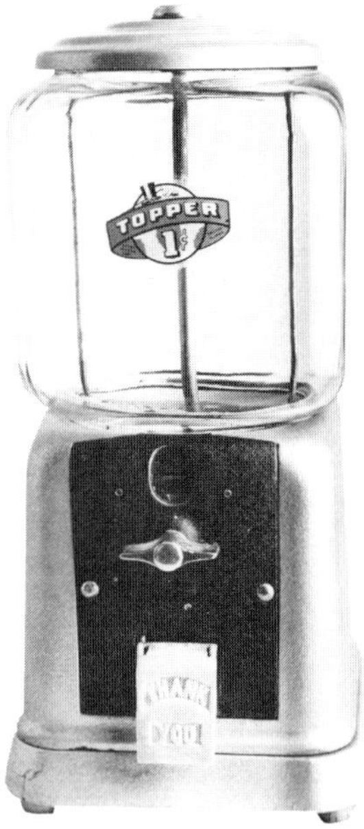

MODEL V Victor changed the name of the Challenger to the Esquire in 1940 and then to the Model V in 1941. During its Esquire incarnation it lost its four feet, however, there was no change in the machine when it became the Model V. The prewar model was cast iron and postwar models were made of different alloys and were 5 lbs. lighter. Cylindrical globes were the first type used followed by the square globe or the durable steel cabinet with a plastic front. **Victor Vending Corp.** 1941

UNIVERSAL This post war bulk vendor was originally available with the cylindrical globe and several years later a square globe was offered. **Victor Vending Corp.** 1946

TOPPER The Topper was the mainstay of the Victor line. This model has the square glass globe but was available with other tops or cabinets as can be seen in the other pictures. **Victor Vending Corp.** c. 1950

L. C. TOPPER The "L. C." stands for "large capacity." The large plastic globe held 1,175 pieces of a 100 count gum (different than today's 100 count). **Victor Vending Corp.** c. 1950's

TOPPER DELUXE The full cabinet top was made out of metal and plastic and was meant to cut down on breakage. The Topper Deluxe came in brilliant colors and was trimmed in chrome. **Victor Vending Corp.** c. 1950

TOPPER HALF CABINET This machine was available as a Topper or Topper Deluxe. The 7 – 8 lb. globe is a combination of steel and lucite. The base remains the same. **Victor Vending Corp.** c. 1950

PINBALL The machine had a pinball playing field based on a sports theme. It came in a golf, baseball, football and basketball version. **Victor Vending Corp.** c. 1950's

TOPPER This Topper has a plastic globe. The globe can be used upside down by changing the collar on the body and by changing the square lid to a round one. **Victor Vending Corp.** c. 1950

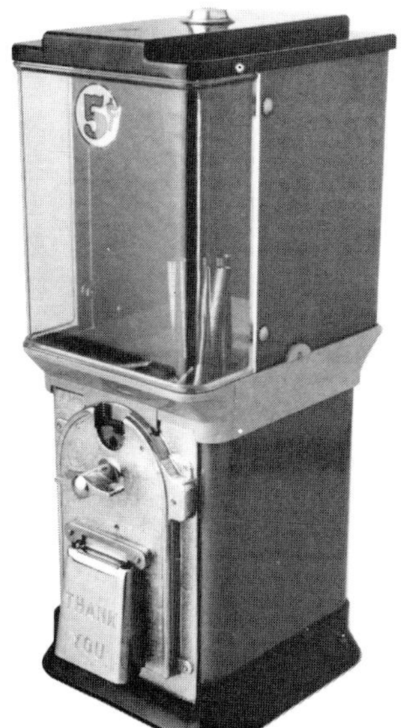

TOPPER DELUXE This Topper Deluxe has a larger capacity, extended half metal cabinet. The base remains the same. **Victor Vending Corp.**

TOPPER DELUXE MUSIC TOO The customer would hear a tune played for up to thirty seconds and receive his merchandise as well. **Victor Vending Corp.**

BABY GRAND DELUXE This oak cased machine had a "Project–O–View" shatterproof window of transparent lucite. It has separate locks for the merchandise and the money. **Victor Vending Corp.** 1951

BABY GRAND The extension on the top of this machine provides for extra capacity and privided product visibility from any angle. It was available in penny, nickel or dime models. **Victor Vending Corp.**

SUPER V Many collectors call this the Grandad. It was a capsule vendor that came in penny, nickel and dime models and had a two-tone solid oak case. **Victor Vending Corp.** 1954

VICTOR 2000 Guess how many gumballs this machine holds—2000 of the 100 count gumballs (different from today's 100 count).The cabinet is oak and the globe is plastic. **Victor Vending Corp.**

VICTOR CARD VENDOR A nickel would buy a gumball and a card from this machine. It is a Baby Grand with a card attachment. This is a desirable machine for baseball card and coin op collectors. **Victor Vending Corp.**

BASEBALL Note the names of the baseball teams on the glass: Brooklyn Dodgers; Cincinnati Red Legs; Washington Nationals; etc. The object was to flip the gumball into one of the holes. The machine also came in a basketball and football version. **Victor Vending Corp.** c. 1950's

FOOTBALL The game continued until you flipped the gumball into one of the holes. The machine also came in a basketball and a baseball version. The sides of the machine were made out of pressed board and had a sheet metal top. **Victor Vending Corp.** c. 1950's

VENDORAMA This is the globe style Vendorama. It is refilled through the top and the cash drawer is removable from the front. Available in penny, nickel, dime and quarter denominations. **Victor Vending Corp.**

VENDORAMA HALF CABINET The half cabinet cost 50¢ more than the all glass globe. The only difference between the two Vendoramas is the globe. **Victor Vending Corp.**

MODEL 88 The Model 88 is one of the newer models that can still be found on routes today. **Victor Vending Corp.**

TOM THUMB The Tom thumb was first made by the Walker – Freeman Co. and later by the Fielding Mfg. Co. The earlier model had a curved, uncovered chute, while the later model had a differently shaped, covered chute. **Walker–Freeman Co., Inc.** c. 1934

WALTER'S BARGAIN VENDOR Walter's Bargain Vendor was available as a 1–2–3 machine or two for a penny. The machine was available in either aluminum or cast iron. **Walter Gum Co.**

MODERNE PEANUT VENDOR This small booth vendor is made of sheet metal with a satin stripe chromium finish. **A. M. Walzer Co.** c. 1935

MODERNE VENDOR (Hershey Vendor) There are two versions of the modernistically designed vendor. The first is the 1¢ Hershey vendor that holds 125 Hershey Bars and is the more common model. The second holds 125 pieces of stick gum. **A. M. Walzer Product** c. 1930's

WESTERN This machine has "Western" embossed on the front flap but it is obvious that it was made by Shipman Manufacturing and Sales Company. It was available in a hot or cold nut version and utilized a pull chain instead of a lever. **Western Co.** 1932

YORK Ford may not have been too happy with this early look–alike of their machine. It has a round globe, smaller than Ford's, embossed with "York." The base looks just like a Ford. The decal reads, "York Vending Co., East Aurora." That is the city where Ford was started. **York Vending Co.** c. 1920's

SWEET BREATHS The Yu-Chu sanitary gate allowed the company to brag about this machine being absolutely dust-proof. The machine fed the gum directly into the hand. These sanitary models were available for gum or Sweet Breath Pellets. **Yu-Chu Co.** pat. June 2, 1925

GUMBALL The Yu-Chu gumball vendor is a pot metal, jar-topped machine. The globe in this picture is one of the jars most frequently seen, although other jar globes are correct. The globe that is pictured on the breath pellet machine is also correct on this machine. **Yu-Chu Co.** pat. June 2, 1925

YU-CHU Actually, this is a Master #77 or Fantail with a Yu-Chu Co. flap. Norris personalized their machines by putting their customer's name on the flap if you gave them a large order. **Yu-Chu Co.** C. 1933

ZENO Like many of the early machines the Zeno gum machine could be acquired free with the purchase of 1200 pieces of assorted Zeno gum. Some of the flavors that were available were: vanilla cream, yucca, lemonade, pineapple and forbidden fruit. Zeno even made gum for Wrigley and was eventually bought out by them. The machine is made of oak and used a clockwork mechanism to deliver the gum. **Zeno Mfg. Co.** pat. July 15, 1902

ZENO Like the other Zeno machines, the domed machine also has a clockwork mechanism. Differing from the others, it dispensed a nickel package of gum and is large for a one column machine.This machine is an unusual example of a domed machine. **Zeno Manufacturing Co.** pat. July 15, 1902

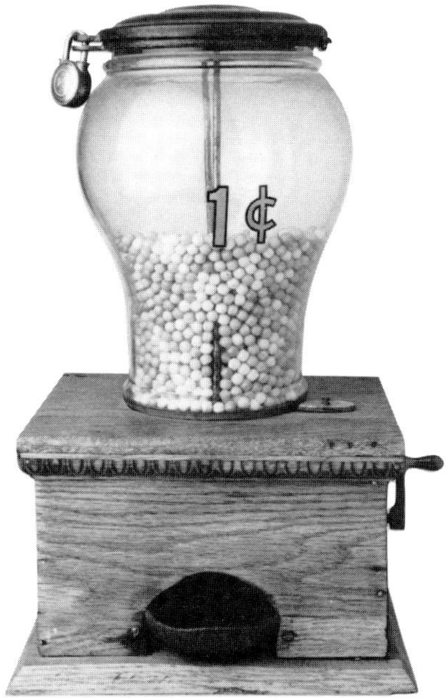

MYSTERY 1 The base of this machine is made out of wood and it utilizes a lantern type globe and a cast iron lid. I would guess it was made in the early part of this century. The manufacturer is unknown.

MYSTERY 2 This aluminum bulk vendor looks very similar to a Doty – Runnels which was made in Kalamazoo, Michigan. However, this machine has a different coin entry and it has feet.

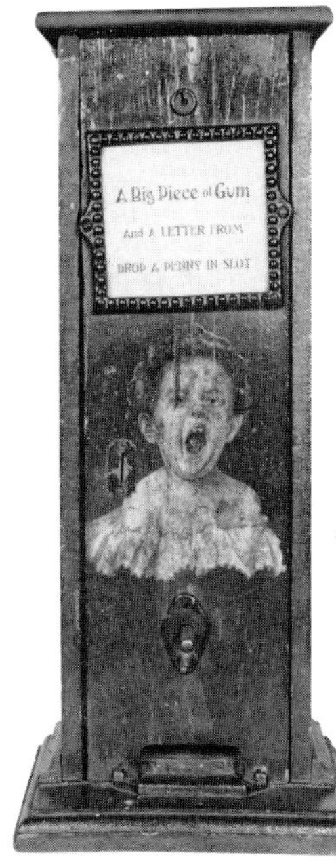

MYSTERY 3 The machine is made entirely of wood and is painted green, with a picture of a little girl on the front. It says on the card, "All for 1¢, A big Piece of Gum And A Letter From Your Sweetheart, Drop A Penny In The Slot." The sides are stenciled with the same message.

MYSTERY 4 The customer got to keep shooting until he put the gumball into the basket. The machine is made of sheet metal with a glass window.

MYSTERY 5 A beveled mirror is on the front of this 2–column vendor housed in a wood casing. The machine sold strawberry or blood orange gum. It reads "J. H. Moore, Canadian Agent" on the card.

MYSTERY 6 The coin drops through a play field which has a background of banners. You always received a gumball and you stood about a 10% chance of getting your coin back. The cheap wood case is speckle–painted.

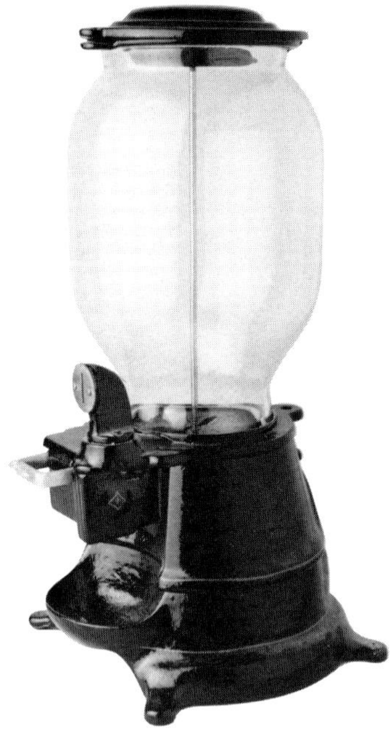

MYSTERY 9 It can't be seen from this angle, but there is a metal divider that goes down the middle of the globe. The operator could put different merchandise in each half. The machine has two coin entries and two levers. This cast iron machine has a "Simpson feel" about it but I don't know who really made it.

MYSTERY 10 I'm not sure this machine is a 100%. The body, periscope coin entry and lid are very similar to a Hance, however, the pull handle and the globe are different.

MYSTERY 11 Not much is known about this machine. The coin slot on top is embossed with a hand pointing to the coin entry and reads, "Penny or Nickel." Note how the coin drops through the globe. Disregard the decal on the globe.

MYSTERY 12 This seems to be a Columbus that an operator has modified. Note the unusual design on the chute flap and the great slug rejecter.

MYSTERY 13 This cast iron vendor dispensed Sweet Chocolate on one side and The Finest Chewing Gum on the other.

MYSTERY 14 The metal globe on this machine looks like a diver's helmet. A French book on coin op referred to this machine as a Columbus. It looks more like a machine from the Ohio Vending Machine Co. of Columbus, Ohio to me. The front flap is missing.

MYSTERY 15 This large machine (15"H × 13"W × 9"D) is made of sheet metal with a crinkle–paint finish. It delivers three gumballs for a penny. Intuition tells me that Bluebird was the manufacturer, but I have no concrete evidence to support my hunch.

MYSTERY 16 A ballerina and a music box sit on top of this 5¢ bulk vendor. They are activated with each pull of the handle. The base is a Silver King look-alike and the globe is a half metal cabinet.

MYSTERY 17 Could this machine be a Simpson? It is chrome over copper which is something Simpson did a lot of. The lid, coin entry, and body all have a feel of being made by Simpson.

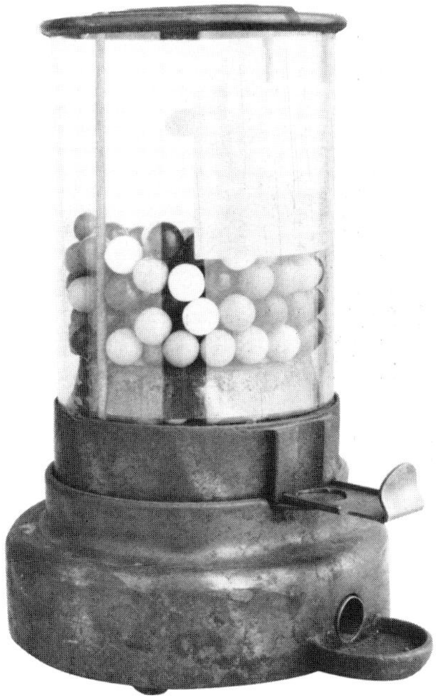

MYSTERY 18 The label is so faded on this early gumball machine that all that is legible is "Chew Digesto Gum" and "Susquehanna, Pa." **Digesto Gum**

MYSTERY 19 Because it has a lantern–type globe, this machine has been called the Lantern Vendor. It is made of cast iron and not much else seems to be known about it.

MYSTERY 21 The drawer where the nuts fall is actually a scoop on this early, primitive bulk vendor. The lower front and side glass panels housing the money in the machine would make this an easy target for thieves.

MYSTERY 22 This chocolate vendor is German. It is a large machine, as it stands about 6 foot tall. The sides have porcelain signs advertising the chocolate.

MYSTERY 23 This is a very early peanut machine with an unusual mechanism. I don't know if the galvanized globe is original or not, however, you can tell the globe is old.

MYSTERY 24 Note how similar this machine looks to the Federal Vending Co. breath pellet machine. However, this machine has the coin entry in the lid and the coin must travel down a chute through the globe to the bottom of the machine.

MYSTERY 25 The machine looks like it was made by Victor Vending Corp. The automaton inside the machine is battery operated. The customer got his merchandise and was able to see the automaton go through his gyrations—all for 5¢.

OLD ADVERTISING

ORBIT CANDY MACHINE.

ONLY PERFECT AUTOMATIC CANDY VENDOR.

Drop a Penny in the Slot, Press the Lever, the Machine Delivers the Candy in the Hand.

Candy Shows in the Machine, and invites Trade Candy Lovers Can't Resist it.

Can't be Cheated.

WILL NOT WORK UNTIL CENT IS DEPOSITED IN THE SLOT.

MONEY ALWAYS IN SIGHT.

Slugs, Tin, Paper, Etc., Etc., go Through.

To Try to Beat it with Spurious Coin is to Pass Counterfeit Money.

Boss of all Vending Machines.

For Drug Stores, Groceries, School Stores and all Busy Places frequented by Women and Children. **Pays for Itself Quick. Always taking in Money.** Put in the Candy; the Machine does the rest.

WRITE FOR SPECIAL PRICES AND FURTHER PARTICULARS.

Perfect Gum Machine

Can't be Manipulated.

Money Always in Sight.

A Piece of Gum for Every Cent put in.

NO FAILURES

1 to 4 Pieces Taken out at once.

Can Use Any Gum Size of Yucatan

Holds 500 Pieces.

It is Passing Counterfeit Money to put Slugs, Tin Tags or Spurious Coins in This Machine.

Nothing but Cents can Work it.

No Gum can be Taken out till Money is in.

☞ If you use this machine you are not tied up to any one kind of gum and can go in the market for cheap goods and make 50 per cent more than with any other machine. This saving will pay for the machine in a very short time.

☞ **Great for Grocery, Drug and School Stores, etc., etc.,** Women and children put their pennies in this machine as soon as they see it * * It takes up no room and is never in the way. * * It don't need watching and can't get out of order.

A Few of These Machines Will Produce a Nice Little Income.

WRITE FOR SPECIAL PRICES AND FURTHER PARTICULARS.

Candy, gum and other things to eat are also profitable when you sell them automatically with

Mills Vending Machines

You can sell peanuts, candy, gum, chocolate, matches, etc., without giving a minute of your time to the business, and yet make a bigger margin of profit than any ordinary storekeeper can make on the same goods. These handsome machines occupy but little space in your place, and they make good money for you while you are attending to your regular duties. They also **keep your customers from going out** when they want to purchase these articles. Why let the confectioner (or some outside slot machine) get the money when it is just as easy to have it in your own till?

Our new Peanut Machine (No. 2) is the only successful sanitary Peanut Machine ever made, and each of the other machines here illustrated is the best of its kind. These machines are

Standard the World Over

No. 1. Glass Globe Peanut Vending Machine gets 35 cents per pound for peanuts that cost 6c to 12c. Net profit on **one machine,** $4 to $10 per week.

No. 2. Our new **Sanitary Peanut Machine,** filled from top; dust-proof sanitary cup; mechanism absolutely protected against salt.

Nos. 3 and 4. Two-slot and one-slot machines for either gum, chocolate or matches. Ask any operator what these machines make.

No. 5. A novel machine for selling mixed candy, **The Mills Windmill Candy Machine.** Arms of windmill revolve when coin is deposited, pick up candy and throw it out through chute. Sells peanuts also. Takes in as high as $10 per day.

No. 1

No. 2

No. 3

No. 4

No. 5

The Mills Peanut and Candy Vending Machines

THE COMBINATION ORBIT

is a double machine intended to vend salted peanuts from one side and candy from the other. In mechanical detail and finish it is similar to the Orbit No. 3, and is supplied, when desired, mounted on a highly ornamental steel pedestal. The earning capacity of the Combination Orbit is practically double that of a single machine, while its cost is but little more.

THE ORBIT No. 3

is similar in operating mechanism to the Orbit, but is enclosed in a beautiful oak case, highly finished and trimmed in either oxidized copper or nickel. The machine is supplied either with or without pedestal, and may be used for vending either peanuts or small candies. The case being of wood is not affected by dampness and may be used in coast cities with entire satisfaction.

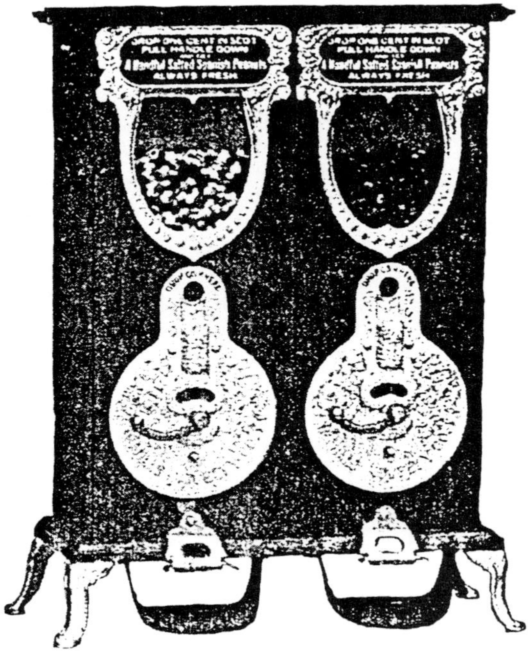

Mills Combination Orbit

Weight, with stand, boxed for shipment, 90 lbs.
Weight, without stand, boxed for shipment, 65 lbs

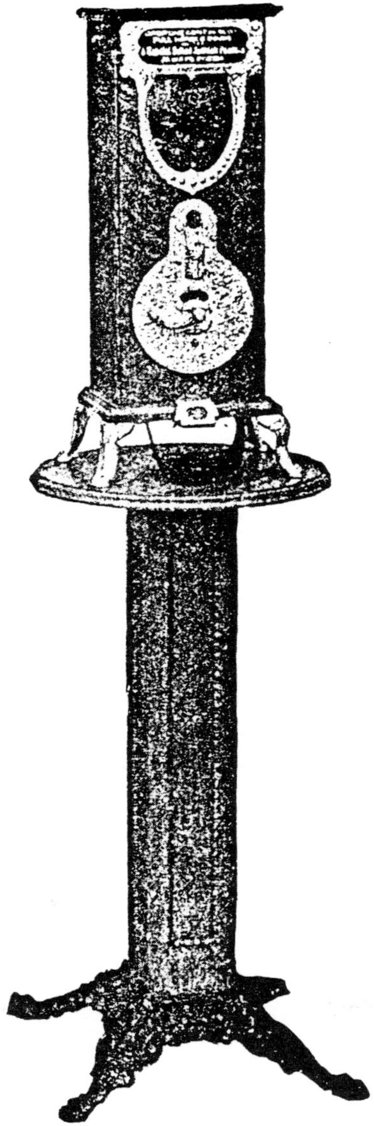

Mills Orbit No. 3

Weight, with stand, boxed for shipment, 65 lbs
Weight, without stand, boxed for shipment, 40 lbs

Star Peanut Machine

This machine has been built to meet the demand for a high class reliable vender. It's far superior to the old style cheaply made iron machine which can be operated by a wire or knife blade and delivers the contents into a rusty iron cup.

The Star is a handsome machine in keeping with the finest store fixtures. You can place it where the cheap unattractive machines can never get in.

It has nickel plated case covered by a transparen enamel. Mechanism is solid bronze, will not rust. The pan and swinging cup are white enamel. The cleanest and most sanitary machine on the market. For either pennies or nickels. Throws out slugs, proof against knife blade or wire.

It will vend peanuts, almonds or small candies, and is the only machine that will do this successfully, due to the patent agitator in the globe.

Get the best when you are buying. It is the cheapest in the end.

PRICE

$15

A FEW FACTS **Peanuts cost from 8c to 12c a pound. Vended through this machine a pound brings in 35c - 25c clear profit on each pound. This amounts to dollars in the run of a year. Think it over and add these dollars to your savings account.**

Caille's IMPROVED Ball Gum Vender

Simplicity of Design, Convenient Size and Durable Construction Makes This the Best Ball Gum Vender on the Market

THE idea of selling to the public Gum in the form of balls is by no means a new one but is without doubt the most convenient way and the most popular.

Several devices have been offered to vend this gum but in practically every instance trouble has arisen.

The essentials of a perfectly working machine are: Ease of operation, positive mechanical construction, coin slot which admits only pennies, large gum chamber, water and dust proof and attractive lines. All of the above are found only on the Caille Improved Ball Gum Vender.

Coin Slot Arrangement

This is the most clever device on the market and does its work with absolute accuracy. Only pennies are admitted. Anything of larger size, thicker or thinner or of the nature of washers is rejected.

As a Trade Stimulator

In addition to the simple idea of vending gum there is a possibility of stimulating trade by mixing in a few balls specially marked or colored good for 5, 10, 15 or 20c in trade and redeemed as they are withdrawn by a lucky patron.

Height 12 inches   Width 6½ inches
Weight 10½ pounds

PRICE - - $10.00
(Including 1000 Balls of Gum)
Extra Gum 10 cents per 100 Balls
SPECIAL PRICE IN QUANTITIES

Special Agitator

One of the most troublesome points in the ordinary gum vending machine is that part of the mechanism which delivers the ball of gum. On the Caille Improved we have entirely eliminated all annoyance by installing our SPECIAL AGITATOR, which positively delivers but ONE ball of gum for one coin and prevents the clogging of these gum balls in the globe or gum chute.

The MONEY BOX is underneath the machine and is arranged to cover the mechanical parts so one lock secures both.

GLOBE is six inches high, holding about 500 balls of gum and is surmounted by a metal cap, making the entire outfit waterproof. The gum is put in through this top, so it is not necessary to detach globe.

MECHANISM is simple yet strong and accurate, and is held in place by two screws. Entire front coin slot arrangement is nickel plated and can be removed in a moment's time. Body and top of machine are highly enameled, being baked on, which insures long wear.

WE GUARANTEE the Ball Gum Vender to be the most accurately working device of its kind on the market.

Mills Package Vending Machine

1905 Model

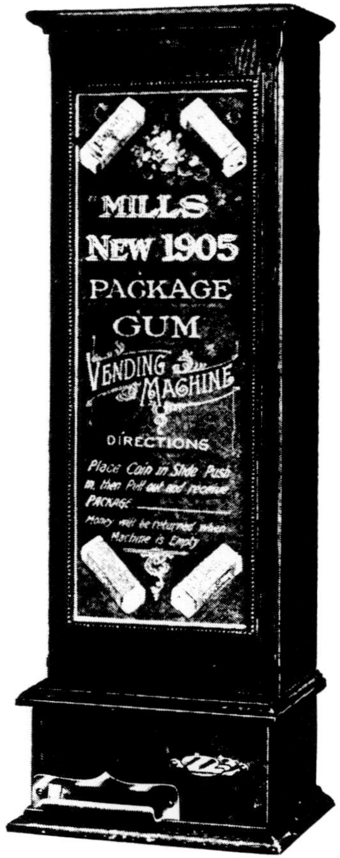

For 5c or 1c

Will vend a package about 3 inches long, 7/8 inch wide and 3/8 inch thick.

21 inches high. Shipping weight, 25 lbs.

Boxed for foreign shipment: Dimensions, 36 x 15 x 12 inches. Weight, 60 lbs.

We make to order Vending Machines for any size of package.

As PREMIUMS to Dealers and MONEY-MAKERS for Operators

Hance Vending Machines Are Supreme

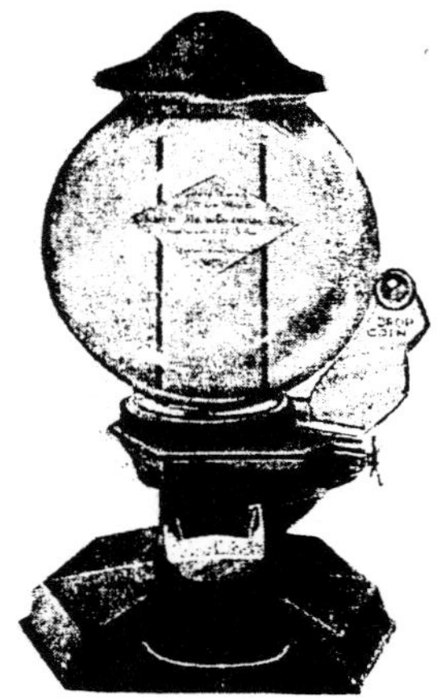

To the Manufacturer: If you produce a line of goods sold in Confectioneries, Restaurants, Drug Stores, at Cigar Stands, etc., it requires no stretch of the imagination to see that if you hold out to your dealers as an inducement, or present them as a reward for patronage, with an article that they can operate permanently at a big margin of profit in their places of business, **your premium will outweigh all others in inducement value.** The merchant whom your advertising, your dealers' helps, or any article of quality premium merchandise failed to influence, will jump at the chance to get one or more Hance Vending Machines to operate in his store by concentrating his purchases with you.

To the Merchant: If you are not so fortunate as to be able to secure Hance Vending Machines as premiums, you can put a few of these "silent clerks" to work for you on a very small investment of capital compared with the big profits they return. You won't have to watch these clerks (save occasionally to fill the containers and relieve them of their earnings); you pay them no salary; you keep no check on them with cash registers and sales slips; you don't have to relieve them for lunch, a "night off" or the ball game. Most of you are pleased to handle lines that net you 5% clear money. Hance Vending Machines usually **net their owners several hundred per cent.** Think it over.

Start your Own Business *now*

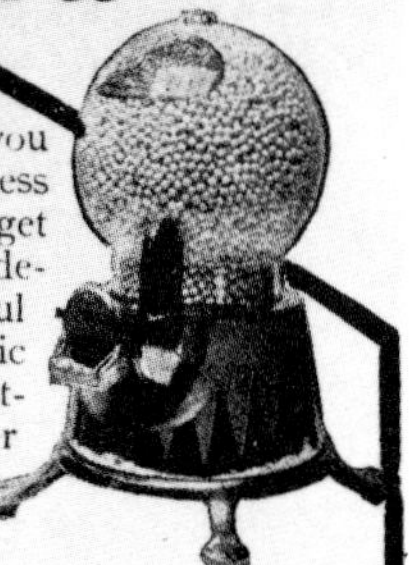

Here is a sure, safe, practical proposition that will put you in business for yourself with less than $10 capital—that will get you *your* share in rapidly developing business of wonderful possibilities. The Automatic vending machine has proved itself a great big money-getter to thousands of successful business men all over the country. They are getting rich because they saw a big opportunity and seized it. The same opportunity with the same brilliant future is offered to *you* right *now*. What will you do?

Also make the best peanut machine on the market. Can be used also for parched corn, salted almonds and small candies

The King Breath Perfume Machine

is the newest, the most effective vending machine yet introduced. It offers a soothing throat-ease and a breath perfume in convenient form and quantity. It is *looked for* when other machines are passed unnoticed. Big profits are yours, if you will reach out and get them.

Mail Us a Postal for Information.

Let us tell you about a trial proposition that will not risk one penny of your money. This investigation will cost you nothing. If you want the profits that belong to *you*—put your name and address on a postal and mail it to us right *now*.

KING VENDING MACHINE CO.,
657 E. Long St., Columbus, Ohio.

Double-Triple BALL VENDOR

NEW, NOVEL PREMIUM FEATURES

Also Made to Sell Salted and Candy Coated Peanuts. Legitimate Everywhere

Sells more gum in a day than the ordinary ball gum machine sells in a month.

Special Offer
One Machine and 3000 balls of gum
$15.00
Best Quality Ball Gum
$20.00
Case of 100 boxes

Made in all flavors and colors.

Special Prices to Jobbers and Operators.

WALTER GUM COMPANY
736-740 Flushing Ave. BROOKLYN, N. Y

Mills Novelty Co.
Coin Operated Machines

37

Little Perfection Nut Vender

Little Wonder Nut Vender

Peanut Machines

A pound of peanuts costs from 6c to 12c a pound according to quality and location. Vended by one of these machines, this pound contains 35 one-cent portions, or a full 25c profit on each pound of nuts. Three pounds can easily be sold a week, hence the machine will yield a profit of 75c a week. Ten machines should *earn an annual profit of* $390. Figure up the income you would receive from 50 or 100 of these excellent little machines.

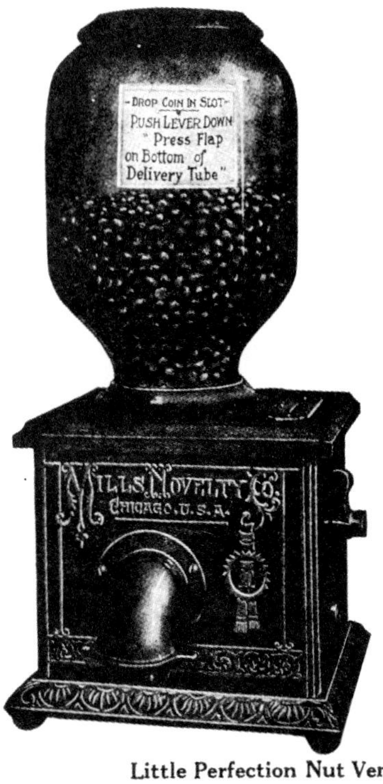

Height	18 in.
Width	8 in.
Depth	8 in.
Ship. Wgt.	12 lbs.

Little Perfection Nut Vender

A *machine* which ought to be on *your* counter, one of the *best* nut venders made. The large clear glass globe displays the salted nut meats most temptingly, thus drawing a large patronage to the machine. The globe is filled from the top and holds a plentiful supply of nuts. Being both water and dirt-proof, the contents are kept always fresh and dry. The dust-proof sanitary cup is another excellent feature. The mechanism is extra strong and absolutely protected from the action of the salt or weather. Strongly built and nicely finished.

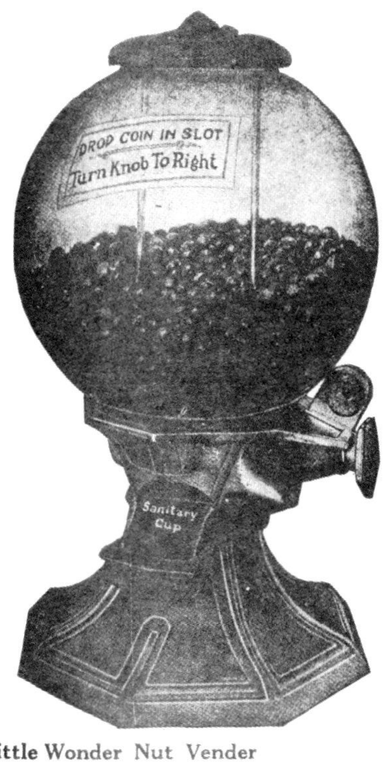

Height	17 in.
Width	8 in.
Depth	8 in.
Ship. Wgt.	5 lbs.

Little Wonder Nut Vender

The Little Wonder Nut Vender, the most sanitary and the lightest and simplest device of the kind ever constructed, *weighs only 3 lbs.* Is made throughout of aluminum, which is practically unbreakable and will not corrode. The large glass, dirt and moisture-proof Globe keeps the nuts all times fresh and dry, and the machine is provided with an adjustable cup which regulates the amount of nuts. Especially good to place out on a percentage basis.

PEANUT VENDER

A Perfect Vending Machine that Will Yield Over 200% Profit

Why Not Let This Silent Salesman Make a Substantial Profit for You?

No. 177H1143.

Sanitary vending machine for automatically selling various kinds of shelled nuts and loose candies. Can be operated anywhere, as it is absolutely legitimate. Gives a good quantity of peanuts for a cent and yet sells them at 36c per pound. A profit of almost 200 per cent. The glass receptacle holds five pounds of salted peanuts. The entire mechanism can be removed and replaced in half a minute. The machine is made of iron; is 20 inches high and 10 inches in diameter at the base, thoroughly enameled both inside and outside and is well ornamented. Each machine has rubber feet and a bracket for fastening to wall or post. The peanuts are delivered by a slight pressure with the fingers, although if desired the machine will deliver the nuts in the cup below by simply removing the sanitary attachment. Each - - - - - - - **$6.75**

Make yourself a sales manager

Increase your income

Put a salesman on every busy corner in your neighborhood. Build up a force that will work for you day and night—that will turn in big profits—that will give you a comfortable surplus over your regular salary—that will develop an income to make you financially independent.

The Kirk Peanut Vending Machine

(Made also for the automatic selling of candy, sweetmeats, etc.)

is the biggest paying—the most easily started and developed proposition ever offered to you. You can start as small as necessary—you can make it as large as your ambition carries you. 1 machine or 100 will earn the enormous dividends of 217%. Do you see the wonderful possibilities of such a proposition to you? Do you realize what a great opportunity is placed in your hands, if you will reach out and grasp it? You can work it along with your present occupation without the least inconvenience. A few minutes' time, a little pleasant work is all that is necessary, until the business grows too big. You can then devote your entire time to it, operate a big business of your own that will be steady, permanent—that will bring enormous profits—that will grow indefinitely.

One of our thousands of testimonials

Am operating 100 Kirk machines in this city, which net me $65.00 a week. Have seen most all of the so-called peanut vending machines, but am glad to state, the Kirk has them all beat to a frazzle. Herewith check for which ship another 100 machines immediately.—S. W. Boyson, 849 Washington St., Boston, Mass.

Let us show you the profit Write today for our illustrated booklet describing the Kirk Machine, showing its durability, beauty, superiority over any other vending machine on the market—its tremendous earning powers. Read the great success that others have had and are having with the machine. Figure out what immense profits are yours, if you want them. Carefully consider our big sample machine proposition that will enable you to prove absolutely the truth of our claims and the wonderful possibilities of the machine. Don't let this proposition get by you or somebody else in your neighborhood will get in ahead of you. Write today—now—mailing us the coupon in the corner. Clip it off before you turn this page and let us tell you the whole story.

KIRK MANUFACTURING COMPANY
1631 Old South Bldg., Boston, Mass.

Kirk Mfg. Co., Boston, Mass.

Gentlemen — Send me without obligation on my part, your booklet on vending machines and your special free sample machine proposition.

Name.........................

....

Address.........................

....

Get this Peanut Vending Machine *Free*

Here's a salesman who will work for you without pay. Instead of asking for a weekly salary, it will *pay you* a clear profit of from $2 to $10 each week.

It's the most simple, safe and successful vending machine ever made. Displays nuts in most attractive fashion. Gathers in hundreds of pennies every day. It is burglar and slug proof—causes you no worry or bother. Never out of order. Always at work for you rain or shine.

All you need to do is to keep your O. I. C. Vender supplied with nuts and collect the money. The machine will do the rest. Each vender holds 8 lbs. of nuts—320 pennies' worth—$2.24 clear profit for you. With each machine we send 48 lbs. of the finest quality blanched salted peanuts. They are the cleanest and most hunger-inciting of nuts—the taste-like-more kind. Your vender will dispose of them all in from two to three weeks at a profit of $13.44.

To those who wish to start a profitable business, we offer for a few days, one O. I. C. Peanut Vender at $12.50, less than the profit you will realize from the sale of your first order of peanuts, so that the O. I. C. Vender actually makes you a present of itself within a very few weeks.

O. I. C. owners with 20 to 100 machines never make less than $30 to $100 per week. It's a big field. Only recently we got an interesting letter from Geo. C. Borreson, Pine Bluff, Ark. Read what he says at the right.

Don't miss our special offer. Start a business that puts itself on its feet and brings you a steady profit without work, or worry. Cut out the coupon at the right and send with your order today. Send bank draft, post office or express order.

O. I. C. COMPANY,
1372 Unity Building, Chicago.

This Coupon Saves You Money I wish to take advantage of your special price offer. Enclosed is $18.26—$12.50 for one O. I. C. Peanut Vender and $5.76 for 48 lbs. of your celebrated blanched salted peanuts.

Name.........................

Address.........................

Town.........................

State.........................

Ship by

Express......

Freight

......

$2.00 a Day From One Machine

"Enclosed please find money order for $3.00 for which send me by express one 25-pound pail of blanched peanuts like the sample you sent me. They were crisp and fresh and very satisfactory.

"I have had the machine up three days now and it has averaged $2.00 a day, which, if it keeps it up, will be very pleasing to me.

"I expect to place more of the machines and will probably send you an order for more this week.

"Hoping to receive the peanuts promptly, I am, Very truly —GEO. C. BORRESON, 111 Main Street, Pine Bluff, Ark."

Gum Vending Machine and Gums

177N2050. Everybody chews gum. Someone must sell it. Why not you? Your profit is enormous. You can buy 100 sticks for 45c, sells for $1.00. Your profit is over 100 per cent. Constructed same as our match venders and in appearance the most attractive of them all. The greatest penny catcher on the counter. A ready silent counter salesman. Should you have this machine on your counter for one week you never would be without it as

long as you are in business. These machines are made of heavy steel, nicely japanned and stenciled. Strong, no delicate or complicated machinery to get out of order. Sold with a positive guarantee. Has rubber feet, suitable for any counter, also fitted with bracket to hang on wall. The ready silent counter salesman—the greatest penny catcher of all. Size 6x6x12. Each vender holds 100 sticks gum. Each, **$3.25**

"Master" Target Practice

"MASTER" MINTS For BELL OPERATORS $10.25 Per Thousand Packages

High Grade BALL GUM Delivered $18.00 per case of 10,000 Balls

HE WHO OPERATES "MASTERS" HAS NO COMPETITION.

Our improved 1925 "MASTER" PEANUT and BALL GUM MACHINE will eliminate steel washers and slugs; with Universal vender, vends either peanuts or ball gum.

Our improved PISTOL TARGET PRACTICE MACHINE with gum vender is the greatest money maker for the merchant and operator ever produced. Write for New Catalog.

MANUFACTURED ONLY BY

THE NORRIS MANUFACTURING CO. 553 Nicholas Street, COLUMBUS, OHIO

The 1-2-3 Machine

Mr. Operator---

NO MAGICIAN'S TRUNK EVER HELD SUCH A WONDER AS THIS—

THE NEW BOWLING ALLEY GUM VENDING MACHINE
Complete
With Lobby Display Stand

IT WAS THE SENSATION OF THE "CHICAGO SHOW"

And Is Gaining, Gaining, Gaining, in Popularity.

Special Pre-Sales Offer To Operators Only—

About April 20th this New Number will be offered to the public at large through our Specialty Salesmen. At that time regular prices will prevail.

As a *pre-sale proposition* to operators only, until April 20th, we offer in lots of 5 machines or more at factory prices without salesman commission.

WRITE OR WIRE FOR PRICES

and be first in your territory to operate this ACE OF VENDING MACHINES.

HONEY DEW GUM COMPANY

Manufacturers and Distributors

1322 Main Street — Kansas City, Missouri

PENNY PROFIT PRODUCERS

A NEW GAME OF SKILL

Rewards and Merchandise Paid Out Automatically

A BIG SUCCESS

Made In Four Styles

Write for Description and Prices

THE NORRIS MFG. CO.,
Manufacturers of Vending Machines
1424 East Cherry St., COLUMBUS, OHIO

The "Silent Iron Salesman"
IS THE
ORIGINAL NICKEL GETTER

THE BEST MACHINE FOR OPERATORS

because the cash box is on the side and large enough to hold 1,200 nickels, that can be taken out without turning the machine over—eliminating chances of breaking the glass globe, showcase or machine.

This Is a Fascinating Vending Machine

which costs a nickel to play. Each ball of Gum has a hole drilled through the center, containing a number which indicates the prize winners as listed on card furnished with every machine. The "Iron Salesman" sells your own merchandise, and can be operated in most towns.

$275.00 PER MONTH NET PROFIT

can easily be made with 10 machines if properly placed. The "Iron Salesman" holds 1,200 balls of gum and $60.00 is realized from every filling. Every "Iron Salesman" placed in live spots will make more real money for you than 25 penny machines.

Come on, you hustling operators, and get into the nickel class, where all the big money is made. Write for special prices for quantity buyers.

Ad Lee Novelty Co. (Not Inc.)
185 North Michigan Avenue, - CHICAGO

UNIVERSALS

ARE GOOD MACHINES

MODEL A
1-2-3 2 Balls for 1c
or 1 Ball for 1c.

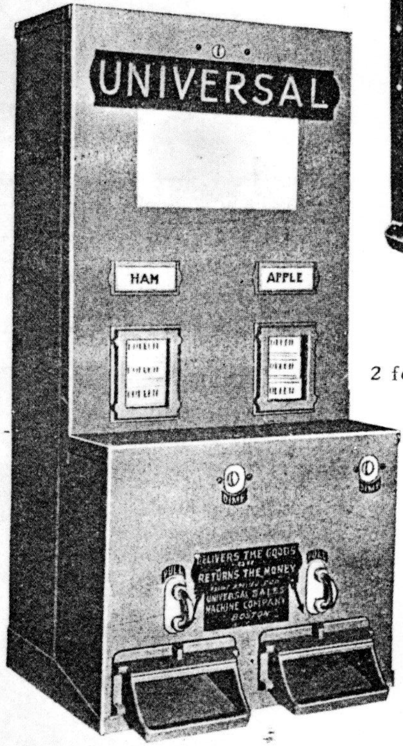

MODEL B
1-2-3
2 for 1c, or 1 for 1c.

MODEL E
Vends anything packed in carton
4x4x¾ to 1¼ inch.

MODEL C
1-2-3 2 for 1c or
1 for 1c.

Chicago Agent,
CHARLES W. HOFFMAN,
6244 Carpenter Street.

MODEL D
Vends ⅞ to 1
inch ball.

UNIVERSAL SALES MACHINE CO.

104 Portland Street :: Boston, Mass.

Model B
1-2-3
2 for 1 cent
or 1 for 1 cent

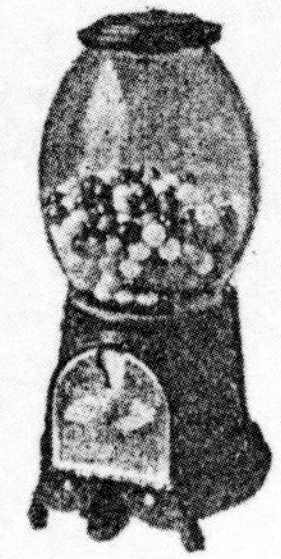

Model C
1-2-3
2 for 1 cent
or 1 for 1 cent

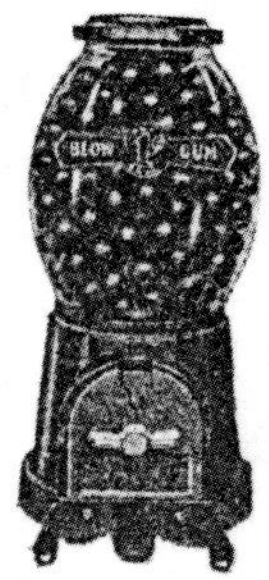

Model D
Vends ⅞ to 1
in. Ball
1 for 1 cent

Model F
Vends anything in bulk.
Has adjustable measure.
Rejects slugs or other foreign
material. Will not clog or jam.
Send for circular.

Model E
Vends anything; packed
in carton
4x4x¾ to
1¼ inch

REMINGTON SERVICE MACHINES, INC.

Factory and General Sales Office:

MAIN STREET, ILION, NEW YORK

Subsidiary of Remington Arms Co., Inc., Makers of Remington Firearms, Ammunition and Cutlery

Remington
SERVICE - MACHINES

BLUE BIRD

"An Institution"

Designers
Manufacturers

SPECIALISTS FOR OVER TEN YEARS IN THE MANUFACTURE OF COIN-CONTROLLED MACHINES

We will design and build any style vending machine to order on contract.

SOME OF THE POPULAR BLUE BIRD MONEY-MAKERS:

Baby Vender
Baby Bell
"Try-It" Ball Gum Vender
Automatic Gum Target
Perfume Vender
"Filling Station"
Package Vender
Cigarette Vender
"Big Boy" Gum Machine
Ball Gum Machine
1-2-3 Gum Machine
Peanut Confection Machine
Mogul Machine
"Bowling Alley" Machine
"Tickle" Machine
Multiple Unit Automatic Merchandisers

You are invited to visit our exhibit in Booth 31, Coin Machine Manufacturers' Exposition, Statler Hotel, Detroit, Michigan, February 25-28, 1929.

BLUE BIRD PRODUCTS CO.

9th & Campbell Sts. Kansas City, Mo.

IT'S NEW! Be the First

To Grab It In Your Locality

THE AUTOMATIC CANDY JAR

(Patent Applied For)

An old and proven money-maker in a new and more beautiful form. Mechanism time-tried and proven. Factory experience of half million completed machines back of it.

A Big Hit at the Chicago Convention

Adopted at Sight by the Watling Mfg. Co., Chicago, for Exclusive Use on Their Scales

It's Beautiful!!

Singly, In Pairs Or Groups

Its Good Looks Get It the Very Finest Locations

BETTER DISPLAY—And display sells merchandise.
BETTER BALANCE—Weight all centered on base.
BETTER SECURITY—Concealed padlock under cover.
BETTER EARNINGS—It's beauty attracts more patronage.

Use It In Groups and Treble the Value of Each Location

Models for Ball Gum, Confections, Peanuts

Salesmen, jobbers, operators: Write for details and quantity prices.
Chicago Distributors: Watling Mfg. Co., 701-7 W. Washington Blvd.

HONEY DEW GUM COMPANY

905 Campbell Street Kansas City, Missouri

$2.00 each

Model G

Flat Gum Machine sells wrapped penny slab gum—Gum 35c per 100 pieces.

Model S

Flat Chocolate Machine — Sells wrapped penny tablet chocolate — Tablets 40c per 100 pieces.

All prices — net cash with order, F.O.B. Newark, N. J.

SUCH VALUE

Grab them while they're Here Boys—There's only 1,000 left. These Machines cost over $5.00 to Build—A Bargain, if there ever was one!

THE NATIONAL GUM CO., Inc.

42-44 Spring Street Newark, N. J.

The New Improved

STANLEY VENDOR

for Peanuts, Ball Gum and Marbles. Also for Pumpkin Seeds.

The Stanley machines are all made of finest grade malleable iron with aluminum handles. Enameled in high polished vermillion color.

We also manufacture machines for any other kind of merchandise and shall be pleased to send you full details of all our machines on request.

Silent Sales Vending Co.

715 Green St., Phila., Pa.

10c THE SILENT SALESMAN 5c

contains 1,200 Balls Finest Quality Chewing Gum. Every Ball Perfectly **Wrapped** in Bright, Beautiful Tin-Foil. All **Colors**—Red, Blue, Green, Gold, Purple, Silver. No Soiled or Unclean Gum. No Dirt, No Dust, No Germs. Clean, Sanitary, Sweet, Fragrant. Every Ball Has Concealed Number Wrapped Under the Tin-Foil. **No Boring and Soiling** of Gum or Numbers Sticking in Bore Due to Hot or Damp Weather. Tin-Foil **Also Prevents Melting** or Sticking of Candy Coating on Gum. Over 150 Numbers Draw Trade Premiums—10c to $2.50. Every Set of Balls Takes in $60.00 and Does it Fast. Some Locations Average $50.00 Per Week. May Be Operated Where All Other Machines Have Been Prohibited. Not a Gambling Machine—Only a Merchandise Trade Booster and Stimulator.

The "Sanichu" Gum Vender is the newest, most up-to-date, attractive machine of any kind that has ever been offered. Requires no Cashier. Delivers its own Merchandise. Always Works. Never Out of Order. And how it does gather the nickels!

First load of Gum pays a nice profit in addition to paying for entire Machine and load of Gum.

"THE SANICHU"

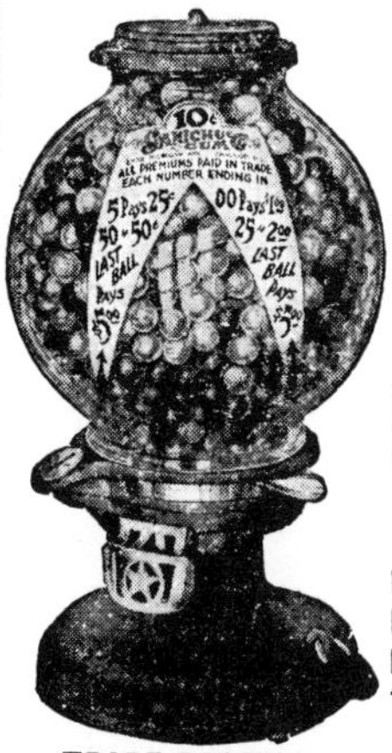

TRADE BOOSTER

Operators— No matter what machines you may have you should investigate and be the first in your field and get exclusive control of the Sanichu, as the Public will buy only Sanichu Gum when it is offered. Every Machine placed will earn $1.00 per day. Special and attractive Terms and Prices to you and exclusive Territory.

Retail Cigar Stores, Confectioneries, Saloons, Billiard Halls and Country Stores— Our Machine will double your profits. Sells your own merchandise. No fancy, overpriced premiums. Trade checks when desired.

Write us for prices and particulars.

Price: Dime or Nickel Machine, Machine and 1,200 Balls Gum, $30.00

Gum for Reloading: 1,200 Balls, $12.00

Salesmen— We want full time or side-line men in every County and State in the Union.

Liberal Commissions and Renewals on all resales of Gum. 100 machines sold will yield $300.00 monthly renewals. One side-line man averaging $20.00 daily. No sample necessary—only Photograph.

Candy, Cigar, Grocery, Soft Drink, Brewery, Sales Board and Specialty Novelty Men, communicate.

Exclusive territory.

This same Machine, for penny play, and 1,500 Balls Gum, in assorted colors and flavors, for $15.00. Liberal commissions to Agents. Also MAY BE USED for Peanuts, Pecans, Filberts, etc.

Ball Gum, assorted, 100 boxes to case, 100 balls to box.

Per case, - - $33.50

1,200 balls, assorted, 4.50

F. O. B. Chicago.

5c THE SANICHU GUM COMPANY 10c

3624 COTTAGE GROVE, CHICAGO, ILL.

OHIO

THE MATES

Model 1

Peanut

Machines are made of of Cast Iron Enameled (Red). Mechanism is made of Aluminum (Non-Rustable). Built to last and give prompt and accurate service, most beautiful and lowest priced machines in existence. Machines can be taken apart without disturbing merchandise in globe. Wall brackets for machines can be furnished at small additional cost.

ORDER NOW

Model 2

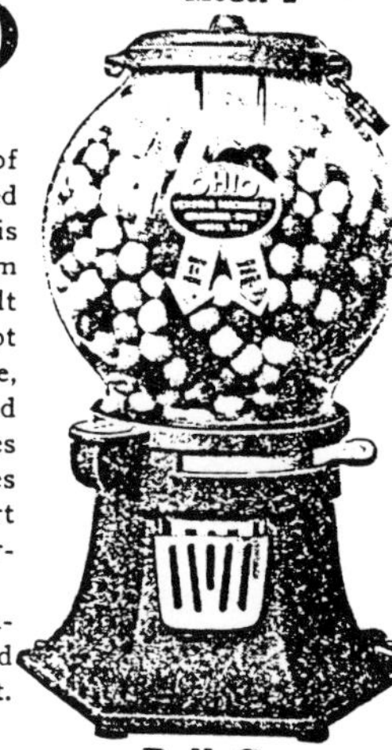

Ball Gum

A TEST WILL CONVINCE

OHIO VENDING MACHINE COMPANY

715 Gustavus Lane — Columbus, Ohio, U. S. A.

ADVANCE VENDOLA

Greatest Money Maker of All Time

The Aristocrat of Vending Machines

Takes the Place of a Candy Stand.

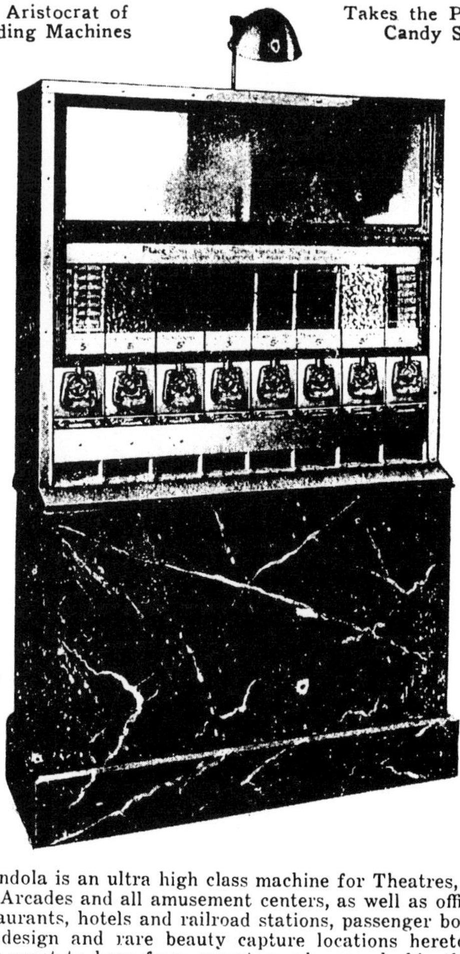

The Vendola is an ultra high class machine for Theatres, Ball Rooms, Cabarets, Arcades and all amusement centers, as well as office buildings, large restaurants, hotels and railroad stations, passenger boats, etc. It's exclusive design and rare beauty capture locations heretofore impossible. We want to hear from operators who can do big things. Write nearest office for information and prices.

See the Vendola at New York Exposition, Feb. 21-24

Rooms 23 and 24.

Advance Machine Co. 4641-47 Ravenswood Ave., Chicago

Advance Vending Co. 554 10th Avenue, New York

A SENSATION!

TRADE MARK

CHEWING BLUE RIBBON GUM

A DELICIOUS CHEW AND PREMIUMS TOO!

The only machine selling A FULL SIZE stick of gum for one cent.

Practically 100 per cent slug proof.

No globes to break! Suitable for indoor or outdoor locations.

Free parts for the life of the machine.

Constructed of sheet metal and brass.

Weighs only 15 lbs.

SOLD ONLY THRU

BLUE RIBBON MACHINES

The wrappers of Blue Ribbon Chewing Gum are redeemable for valuable premiums.

Free Premium catalogs and advertising matter furnished to our agents.

Exclusive operating rights Granted.

We offer operators an opportunity to build up a legitimate business without competition!

Write or Wire Today for our Proposition

"A Delicious Chew and Premiums Too!"

BLUE RIBBON GUM COMPANY, Inc.

28 DODWORTH STREET — BROOKLYN, N. Y.

The California Almond Vending Machine

The Most Beautiful Vending Machine In The World

A High Class Machine For High Class Locations

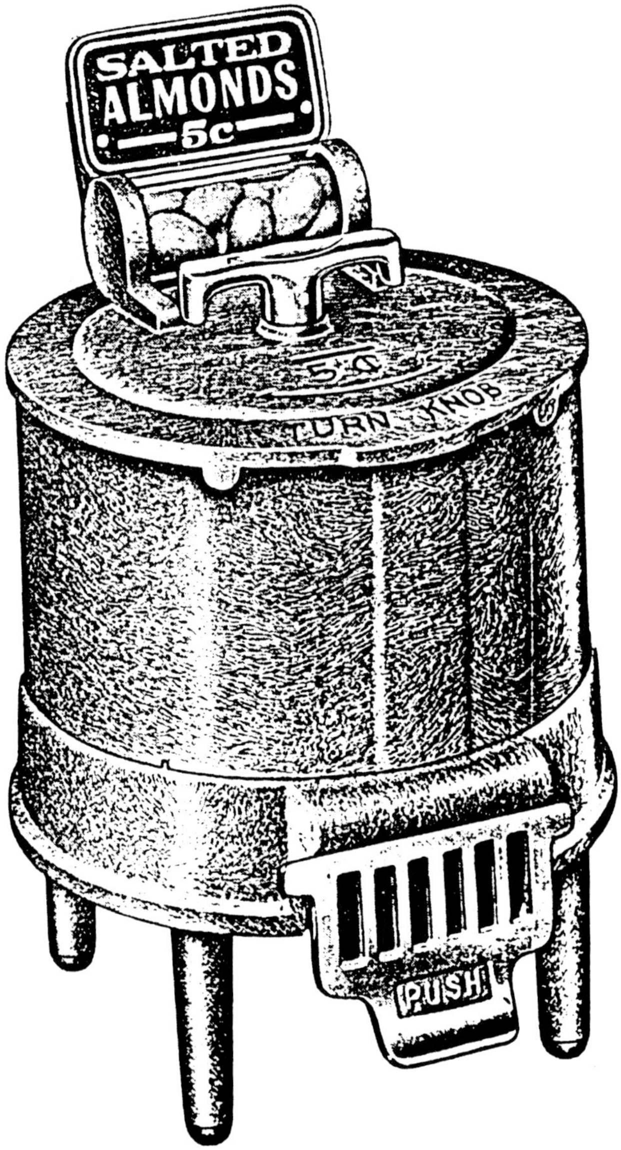

Hard baked enamel finish—in all colors and many artistic designs.

——o——

No shutters or trap doors to cut up nuts. Entire compartments empty when operated.

——o——

Most successful nut machine ever devised.

——o——

Will vend any kind of nuts in the same price class.

Height 9 in. overall Diameter 6½ Weight 7 lbs. Five cent play.

——o——

For the counter or can be placed on wall bracket.

——o——

Shows 300% profit from sale of almonds.

——o——

Capacity ¾ pound.

DESIGNED to meet the requirements of the highest type of trade, this machine is a distinct step in advance. Beautiful in appearance, constructed of the finest materials — bronze, brass and aluminum — it is the ideal vendor for clubs, hotels, recreation halls, drug stores and all places where people gather for business or pleasure.

Shows an enormous profit from the sale of merchandise that cannot be bought through the usual channels for five cents.

This machine excites admiration and enthusiasm wherever it is shown. There are unbelievably high returns in almonds when properly vended, and this machine is proving to be one of the most successful money makers ever placed in the hands of operators.

Three pounds of Supreme California Almonds given free with each machine.
Almonds known the world over.

Sample machine $24.50. One-third cash, balance C.O.D.

Special low prices and terms for quantity lots.
We have a most attractive low priced offer for jobbers.

WRITE FOR PARTICULARS

THE CALIFORNIA ALMOND VENDING MACHINE CO.

1102 DWIGHT WAY, BERKELEY, CALIF.

Actual reproduction of machine and colors. Actual diameter of base, 14 inches. Heighth over all 16 inches.
Get this before your competitor gets it.
Literature and prices furnished on request.

THE ALB COMPANY

BELLEVILLE :: ILLINOIS

The Erie Ball Gum Machine

One of these little money makers in your theatre lobby means a convenience for your patrons and extra profits for yourself. Washerproof and cannot clog.

SPECIAL DEAL
$14.50.

Machine
1,000 Gum
Wall Bracket

REX MFG. CO., - Erie, Pa.

Golden State Peanut Vendor

is of unusual design, pleasingly decorated and highlighted. All metal rust proof construction. Rapid Interchangeable parts permitting the delivery of several types of medchandise—Candied or salted peanuts, jelly beans, ball gum, etc.

Slugs and imperfect coins rejected.

Magazine will hold from 2 to 5½ lbs. of peanuts.

Adjustable to location. Simplified dependable mechanism. Wall brackets at small additional cost.

THIS BETTER MACHINE IS NOW AVAILABLE

Dimensions: 8"x7"x16". Weight 8lbs.

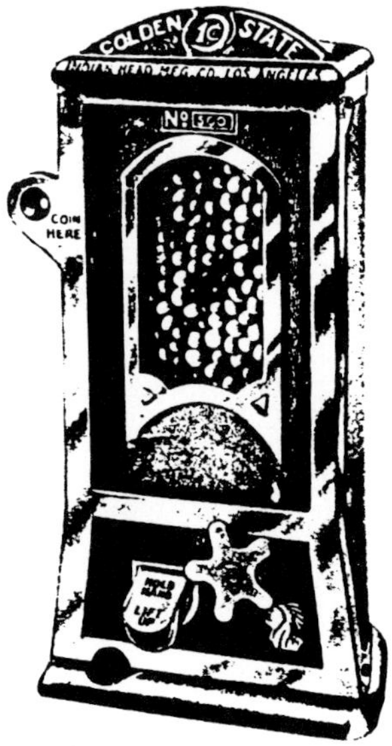

The New "LINCOLN"

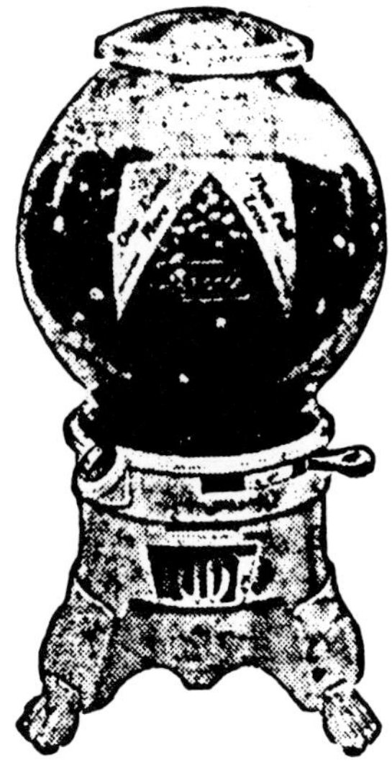

Interchangeable Peanut Ball Gum and Confection Vender. Finished in the New Porceloid Enamel.

Write for Prices.
They will surprise you.

INTERCHANGEABLE

- PENNY CONFECTIONERY -

FOR BALL GUM, CANDY OR PEANUTS

A MACHINE THAT'S NEW WITH A KICK

OCCASIONAL FREE SHOT COIN RETURNED WITH MDSE.

OAK CABINET - BEAUTIFULLY DECORATED

Sample Price, $12.50 F.O.B. Chicago

Write for Quantity Prices

D. GOTTLIEB & CO.

4318-24 W. Chicago Ave. Chicago, Ill.

PROFITS - AT - ONCE

AND WE MEAN JUST THAT

Wrigley's P.K. Penny Package Chewing Gum and Miles Auto Stack is the Winning Combination

Your opportunity to cash in on the tremendous advertising behind this gum vended from the Aristocratic "Miles"

Over 1½ million dollars is spent each week for chewing gum in the U. S.

We have national distributing rights on Wrigley's P. K. Gum through the Miles Autostack.

Over 1000 machines on test last year in Chicago alone.

Holds 600 packages of P. K. Gum. No stacking—just dump them in and lock it up.

The Miles Autostack is the only patentable machine in the world that vends wrapped package gum in bulk without stacking.

A business for you as long as you live—safe, permanent and legal.

One of the most substantial machines ever built and handsomely finished in eye-attracting colors.

A product of Orville L. Miles, mechanical engineer of 35 years standing.

TROUBLE PROOF - CLOG PROOF

FOOL PROOF - WEATHER PROOF

$35.00 per machine (for wall) PEDESTAL $10.00 Extra

For Further Particulars

Address

MILES AUTO STACK CO.

140 S. DEARBORN ST. CHICAGO, ILL.

NEW - APPEALING ATTRACTIVE

The Quality Profit Maker

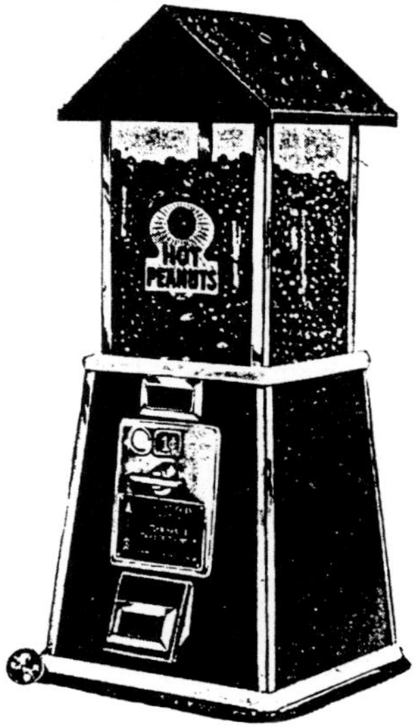

THE AURORA HOT PEANUT VENDOR

You will marvel at this most beautiful, super-quality built machine. And will agree with us that it is the most attractive, substantially built of all hot peanut machines ever produced. Your first order will convince you.

The electric heating element keeps the peanuts fresh, warm and delicious. Warm peanuts, because they taste better—sell faster. This makes it more profitable to every operator. The red light flasher when the current is on, continuously flashes on and off, always attracting customers to the machine. Remember these important facts. Order your machine now, and see how the sales jump over what your ordinary peanut machines can do.

2 STYLES

1c or 5c Machines at the Same Price

You can always change the 1c machine to a 5c machine in 2 minutes by purchasing an extra mechanism. This change can also be made at any time you may desire.

SPECIFICATIONS

SIZE

Height—23″ Width—9¾″

PRICES

Sample Machine $25.00
Lots of 5 .... 23.75

F. O. B. Factory, New York

1/3 Deposit with order, Bal. C. O. D.

OPERATORS — JOBBERS

Write or wire for Large Quantity Prices

AURORA MFG. CORP.

55 West 42nd St.
NEW YORK CITY

2 PLAYS FOR 5c

A Mechanical Triumph

A really efficient coin chute that optionally gives one or two plays for a nickel. Will fit any amusement machine without alteration.

MANUFACTURERS - JOBBERS OPERATORS

Think what a field this opens in the amusement machine industry.

Sample $2.25
12 or More .... 2.00 each

Pat. Pending

Another SENSATION!

Sample Machine **$7.25**

25 Machines **$6.75** Each

TERMS: ⅓ Cash

Balance C.O.D.

Bull Dog Penny Vendors ARE

exceptionally attractive in appearance.

Slug proof — trouble proof — adjustable — adaptable for counter, stand, or bracket.

Choice territories available to live-wire Jobbers.

BULL DOG VENDOR CO.

721 North Main St.
Los Angeles, Calif.

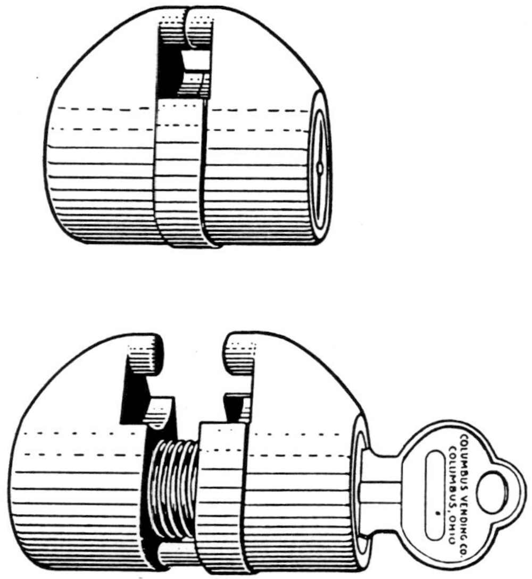

The New COLUMBUS VISE-GRIP LOCK

This new "COLUMBUS VISE-GRIP" Lock has no shackle. It screws together like a vise and has the holding power of a vise. It is of the disc tumbler construction —in short, it is the best lock security yet produced. For further particulars write to

THE COLUMBUS VENDING CO.

2005-2013 E. Main St.
Columbus, Ohio

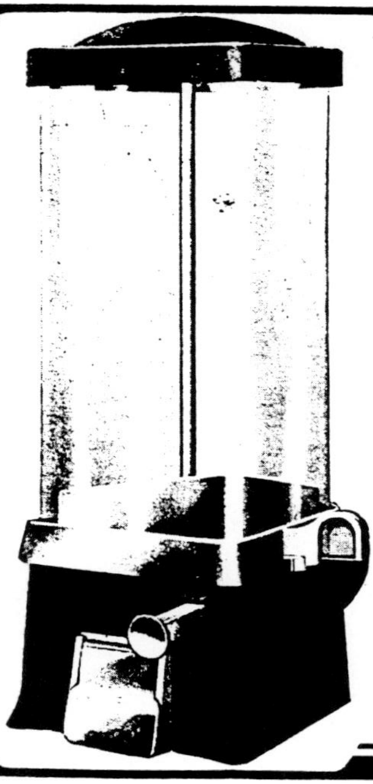

QUALITY OF DESIGN

KALAMAZOO COMBINATION PENNY MACHINE

IMPROVED APPEARANCE

Standard Machine vends Peanuts, Candy, Ball Gum, Mints, Marbles, Seeds, Etc.

Glass Size 4¼ x 4¼ x 9½
Total Weight 7½ Lbs.

Has novel coin control with magnet. Enamel finish, bug-proof case, adjustable delivery. Prices on request.

MANUFACTURED BY

DOTY-RUNNELS CO.
KALAMAZOO MICH.

The Juvenile Merchandise Vender

Built Right - *Looks Right* - *Stays Right*

The Juvenile vender is the latest and best thing out in merchandising machines. From the strong heavily ribbed bottom which is secured by an eleven tumbler lock, to the top cap of the machine will be found sturdiness and beauty in abundance.

Put yourself right with your locations. There is no guess work about Juvenile. Every operator that has put them on location has come back for more.

15 Inches high

These machines have been thoroughly tried out in the toughest locations that can be obtained and they are standing up and giving a wonderful account of themselves.

Get your sample at the special sample price. This price will only last until September 10, when it will be automatically advanced to where it belongs.

Holds 5 pounds of peanuts

Can adjust in one second's time to vend at any amount.

WRITE YOUR OWN TICKET WITH A JUVENILE

SPECIALLY INTRODUCED AT $6.25

Specialty Coin Mach. Bldrs. 617 W. Division St. - CHICAGO, ILL.

Half the Price, Twice the Intake

Money-makers not only for tomorrow, but for the next 50 years.

PETITE VENDERS

Petite Line of Breath Pellet Peanut and Ball Gum Machines

Individual Refined Gets the Money

PEANUT MACHINE
Placed one to each table in taverns, will vend out completely each night. Hold 1¼ lb. nuts. Takes in 75c.

PELLET MACHINE
Holds 2 pounds breath pellets. Takes in $3.00 when emptied. Makes clear profit of $1.00 pound.

BALL GUM MACHINE
Vends either 5/8 in. or ¾ in. ball gum.

Sample machine prices for limited time only. Cash with order.

$4.50 **$4.25** **$4.50**

Specialty Coin Machine Builders

J. H. AUSTIN, Proprietor

617 West Division St. **Chicago, Illinois**

The New

"PERFECTION"

Merchandiser!

"Perfection" results from practice. We've made half a million venders.

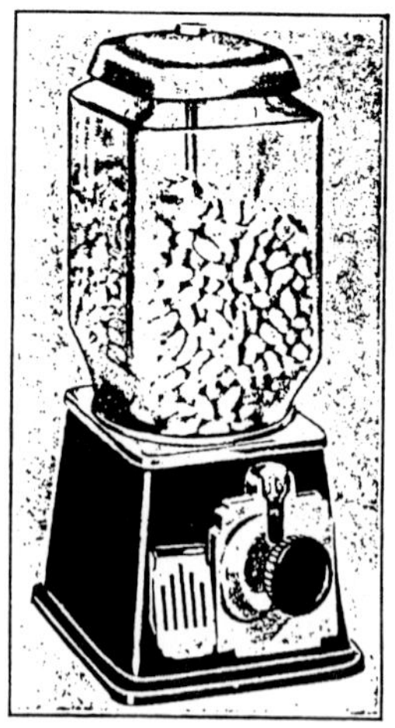

FEATURES:

A brand-new and better coin mechanism—Rubber-sealed globe holds 5 lbs. of merchandise — Vends everything—Slug Proof—Troubleproof — Durable — Streamlined beauty—AND LOW IN PRICE!

DISTRIBUTORS WANTED

SPECIALTY COIN MACHINE BUILDERS
617 W. Division St., Chicago, Ill.

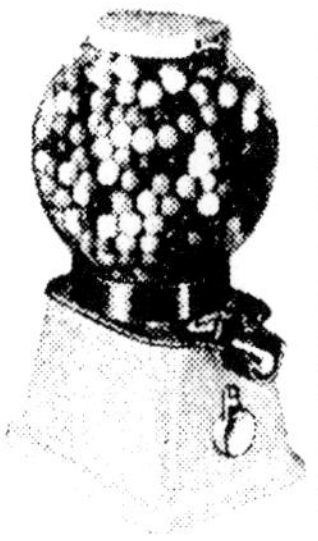

800
BALLS OF GUM IN THE
YELLOW JACKET

Today's livest Merchandise Deal. Precision made, of best material. Eleven tumbler lock keeps collections safe. Holds 800 balls of 1/2" gum or 600 balls of 5/8" gum. Write for prices.

SPECIALTY COIN MACHINE BUILDERS
617 W. DIVISION ST. CHICAGO, ILL.

AVAILABLE NOW!

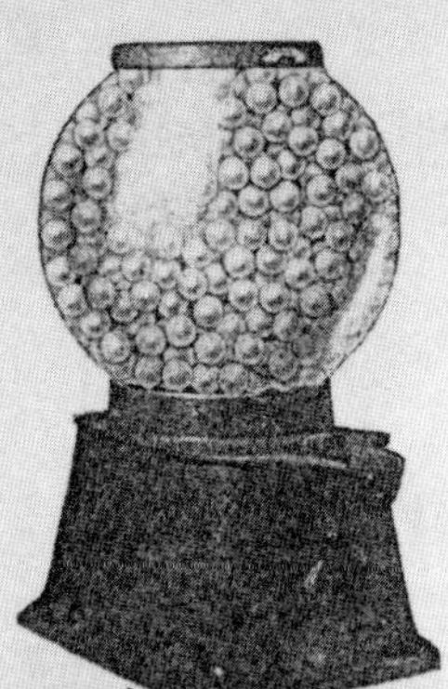

CRAIG'S "999"

- $9.99 Ea. For One or a Hundred
- Mounts on Wall or Pedestal
- Foolproof Delivery
- Baked-On Gloss Finish
- Colors in Rich Red or Brite Green
- Prices F.O.B. Factory
- 1/3 Dep. With Order, Bal. C.O.D.

WRITE NOW! PROTECTED TERRITORIES AVAILABLE.

THE CRAIG COMPANY
6523 MINERVA AVE. CHICAGO 37, ILL.

MODERNIZE
Your Route!

If you operate new up-to-the-minute penny machines you will find more money in the cash box every time you service them. That means more profit for you. Modernly styled Ad-Lee machines only

$6.75 each

Discount on quantity orders.

AD-LEE COMPANY, INC.
825 S. Wabash Chicago, Ill.

BARGAINS IN PERFECTLY RE-CONDITIONED PIN GAMES

(NOTICE: All machines have 5c coin chutes. Remit $1.00 extra per machine for 1c chutes.)

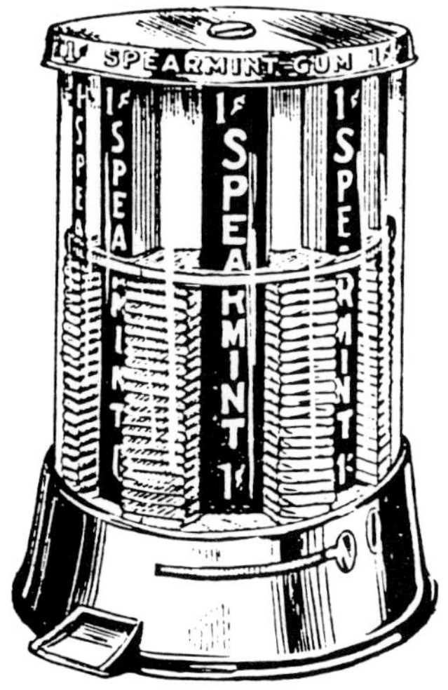

25 Reconditioned Tab Gum Vendors. Capacity: 400 sticks.
$3.50 ea. or $15.00 for 5
We can supply Spearmint Tab Gum at $3.50 per 1000 sticks.

ONLY $15.00
Buys any 4
of the following games

AIRWAY	PROGRESS
BIG BERTHA	LIGHTNING
FLEET	ELECTRO
ACTION	SIGNAL
GOLDEN GATE	DROP KICK
PENNANT	SUBWAY
PIG SKIN	LIVE POWER
REBOUND	FLYING TRAPEZE
GRIDIRON	BEAMLITE

ONLY $20.00
Buys any 4
of the following games

ESQUIRE	KINGS OF THE TURF
CASTLELITE	SUPER "8"
BEACON	TURN TABLE

ONLY $25.00
Buys any 4
of the following games

PUSHOVER	21,000
3 IN LINE	I.O.U. RADIO
ANGLE LITE	SCORE-A-LITE
SENSATION	TRAVEL

MISCELLANEOUS SPECIALS!

10 brand new Totem counter size ticket vending pin games, $10.00 each. Original cost, $25.00 each. This machine gives you a POSITIVE record of awards.

5 Advance 4 column 15c cigarette vendors .... $12.50 ea.
25 Columbus 5c model B peanut vendors .... 3.50 ea.
3 Indoor Striker 1c counter games .... 4.00 ea.

USED MACHINE TERMS: 50% DEPOSIT WITH ORDER, BALANCE C.O.D.

D. ROBBINS & COMPANY

1141-A De Kalb Ave. BROOKLYN, NEW YORK

Counter Gum Vending Machine

Spun Aluminum Finish

Will not Tarnish

SLUG REJECTER

Manufactured by

U. G. GRANDBOIS COMPANY

Kalamazoo, Michigan

Gum Vending Machine

Enamel Finish

SLUG REJECTER

Manufactured by

U. G. GRANDBOIS COMPANY

Kalamazoo, Michigan

A BIG MONEY MAKER

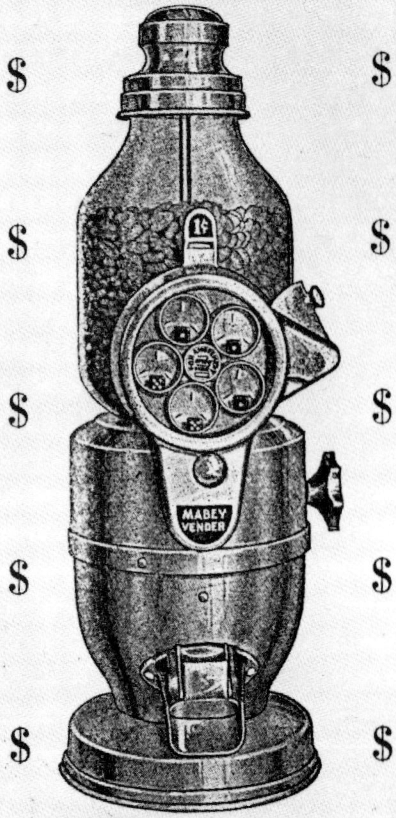

Mabey's "HIGH DICE" Vender

—TRIED and PROVEN—

BY OPERATORS

FLASH GETS THE CASH

7 REASONS WHY

7 SPECIAL FEATURES

An Outstanding Modern Leader

WRITE TODAY FOR FULL DETAILS

Mabey Electric & Mfg. Co.

INDIANAPOLIS - - INDIANA

"FACTORY KING"

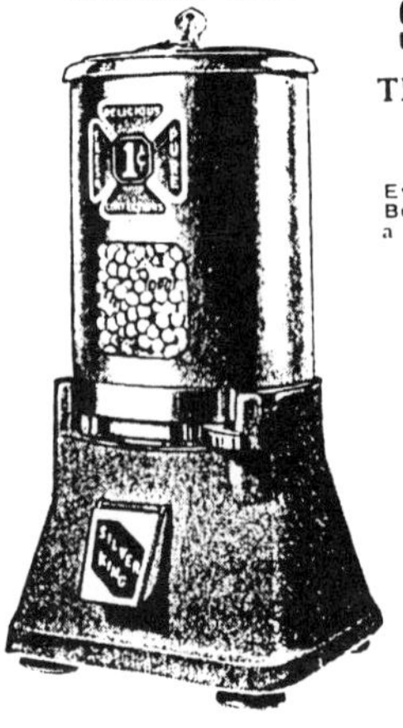

SAVE! ON THESE NEW 1939 MODELS

THE ONE THRIFTY BUY AT BOTH SHOWS

SILVER KINGS

Every operator needs these two SPECIAL VENDORS. Beautiful, modern, trouble free, money makers. There is a "Genuine, Original" SILVER KING for every need.

"FACTORY KING" For tough location

"SILVER QUEEN" For classy location

"SILVER KING" For average location

"SILVER CHARM" For charm location

SPECIAL!

INTRODUCTORY OFFER

One Vendor

10 pounds candy

1 gross charms

$8.45

For limited time only. Don't risk the chance of losing out! Send us your order today, before you forget!

All metal container, display window. For garages, factories, gas stations and tough spots.

Large top! Designed Especially for CHARMS. (Glass protecting metal ring, 25c extra.)

Sample vendor $6.50. Ten @ $5.50 each. F.O.B. Chicago.

AUTOMAT GAMES, 2425 W. Fullerton, Chicago

"SILVER CHARM"

The BELLE ELEVATOR

A REALLY NEW PREMIUM AND MERCHANDISE VENDER

The Belle Elevator

is a merchandising machine unlike anything ever presented before and has a tantalizing operation which will attract both young and old and keep coins dropping continuously and repeatedly into the maichne.

A beautiful front shows a replica of an apartment house and an elevator operated by the player runs up and down the machine. After a coin is inserted the elevator slowly rises and when it reaches the top a bell rings and merchandise and premiums are deposited into the elevator. After receiving its load the elevator slowly descends to the bottom where a door from the elevator opens and delivers the merchandise into a receptacle.

- The Wonder Machine
- Combination Amusement Merchandise and Premiums
- Entirely Different
- Fast Action
- Stupendous Profits
- Sturdily Constructed
- A. B. T. Slot
- Simple Servicing
- A Masterpiece

Special Operator's Price

$29.50

Any Quantity

1/3 Deposit Bal. C.O.D., F.O.B., N.Y.

Stands for Machine

$2.50

SEE YOUR NEAREST JOBBER

or

WRITE FOR COMPLETE INFORMATION

S. & F. SALES COMPANY

1133 Broadway **New York, N. Y.**

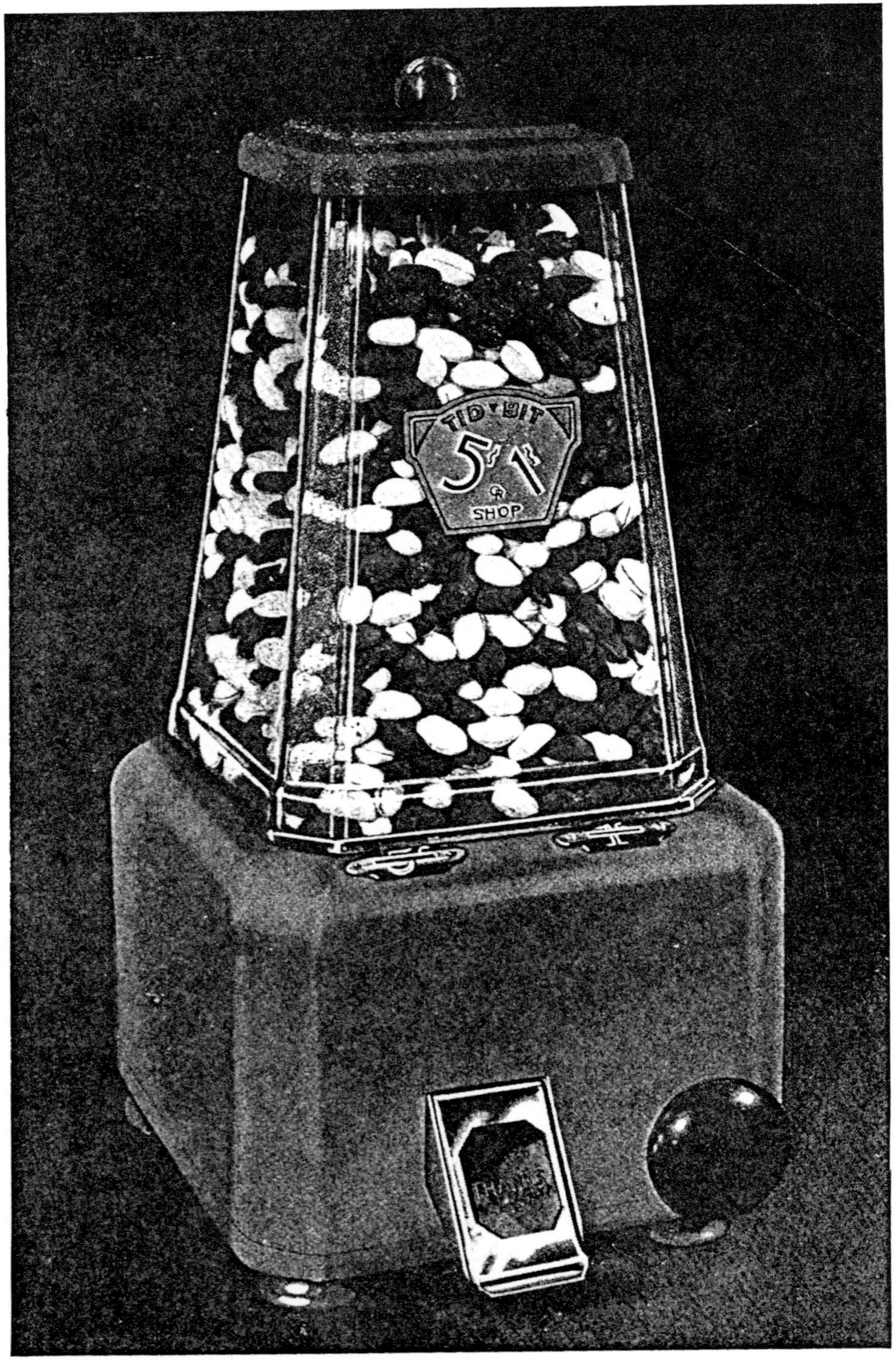

dependable

TID-BIT SHOPS

NEW $35,000 "VENITOR"

Endorsed by Experienced Operators

Ultra Modern, Streamlined Bulk Sales Cabinet. **Extraordinary Beauty - Superior Performance**

●

Exclusive, futuristic, square globe - cannot be twisted.

Bakelite knobs - *first* appearance, fastened securely inside.

Enclosed rejector refuses penny and nickel slugs.

Concealed locks - at no extra charge.

Chinese-tangerine finish - our new trade-color creation.

Mechanism removable without disturbing products - quickly serviced.

Portions adjustable without removing merchandise.

Sturdy 1c and 5c combination mechanism - sizeable parts, easily accessible.

Sells 1c portions and *either* 5 or 6 penny portions for 5c.

Improved delivery assures more uniform portions.

Original "Cartridge Loading" - enables rotating 28 different products.

Lifetime construction - carefully designed, engineered and built of durable materials, properly rust-proofed.

Opens new locations, holds old locations, stimulates activity, increases profits.

A full size machine - 7⅞" x 16⅞" for all around use.

●

Authorized Jobbers being appointed. *Descriptive Circular upon request.* *Wholesale Only*

VENITOR CORPORATION

Tid-Bit Shops *Manufacturers* *Tid-Bit Products*

Title & Trust Building

DETROIT

U. S. A.

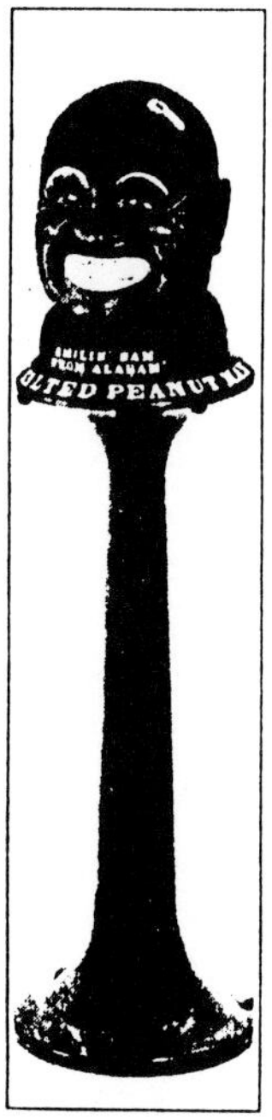

11½ inches High
11 inches Deep
10 inches Wide

He can bring you happiness and prosperity, as he has to hundreds of others.

Price

$41.50

F. O. B. Chicago

Quantity Prices Upon Request

TYPES
White Man — *Colored Man*
Polished Aluminum
Indian — *Chinese*
The only indestructable vendor

"Smilin Sam From Alabam"

Salted Peanut Man

The manufacturer of "SMILIN SAM" Vendor has had over thirty-five years of experience in the Peanut and Vending line.

Paramount Units are not built just for today, but as a foundation for the merchandising machine industry of tomorrow.

The attractive appearance of "SMILIN SAM" makes it desirable for all locations, either interior or exterior of the best stores, clubs, industrial plants etc.

"SMILIN SAM" the Aristocrat of Vendors.
Get your profits from this Silent Salesman

PARAMOUNT VENDING CORPORATION

Madison-La Salle Building

173 WEST MADISON ST. CHICAGO, ILLINOIS

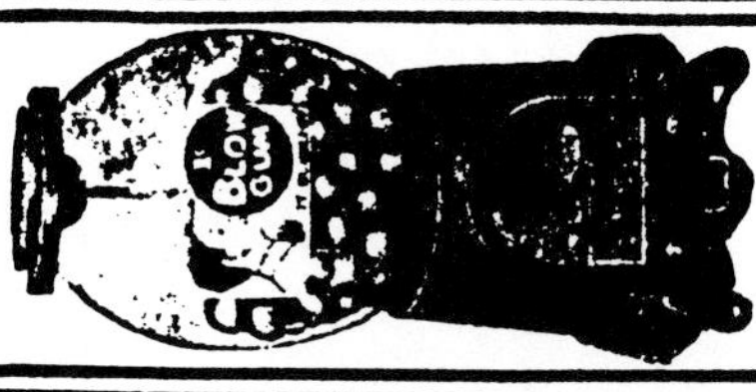

Sell Blow Gum Machines

$75.00 to $200.00 Weekly.
WRITE
CHU-MOAR GUM CO.
WILMINGTON, DEL.

PAY LESS! GET MORE!

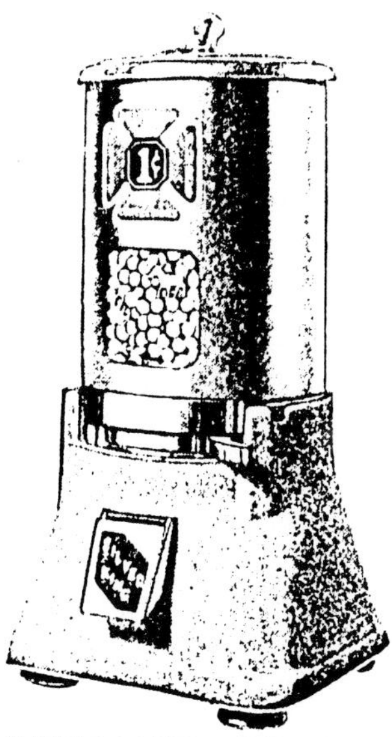

2000 SILVER KINGS A MONTH!!!

THERE'S A REASON!

SPECIAL

"FACTORY KING"

All Metal Container with Display Window For Tough Spots, Garages, Gas Stations, Factories.

Sample $6.50. Ten at $5.50 each.

Special introductory offer. Don't risk the chance of delay! Act now! See any reliable jobber about SILVER CHARM, and FACTORY KING, and SILVER KINGS.

Most Widely Copied Vendor in the World.

Buy only Genuine, Original "SILVER KINGS"

AUTOMAT GAMES 2425 FULLERTON, CHICAGO

1c Dual Merchandiser

"An Operator's Machine"

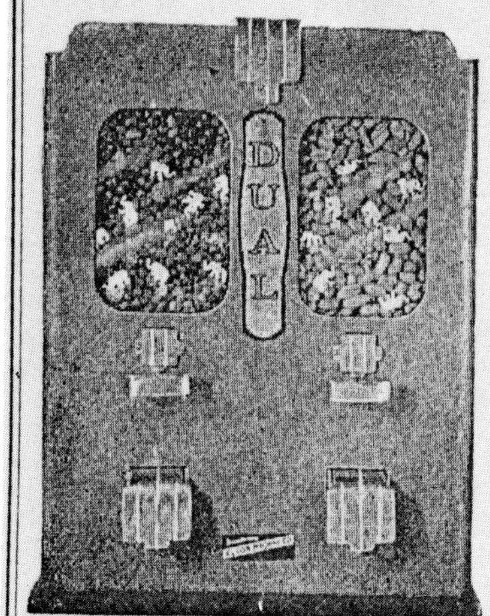

- Sturdy
- Simple Mechanism
- Portions Controlled
- 5 lbs. each compartment
- Vend Nuts, Candies, Toy-Mix, etc.
- No Broken Merchandise
- Location Builder
- Eye Compelling
- Finished all colors
- Actual Size—18" high—14" by 14½"

Operators—Here is the DUAL. Paint your own picture of Dependable Double Profits!

Write for prices and literature today!

NATIONAL DISTRIBUTORS
X. L. COIN MACHINE CO.
1353 Washington Street -:- Boston, Mass.

MICKEY MOUSE AND HIS PALS

New Type VENDING MACHINE

That Will Increase Your Profits 100%

© 1938 WALT DISNEY ENTERPRISES

Here it is! The Biggest News in the vending machine industry. Mickey Mouse and His Pals bring you the "vendor that has everything"—with an assurance of 100% increased earnings. Now ready for Delivery!

CHARMS—

All the lovable Disney movie stars—Real personalities — the charms are reproductions of original hand engraved Disney characters. Available only to users of Mickey Mouse and His Pals Machines.

STAINLESS STEEL TRIM Throughout

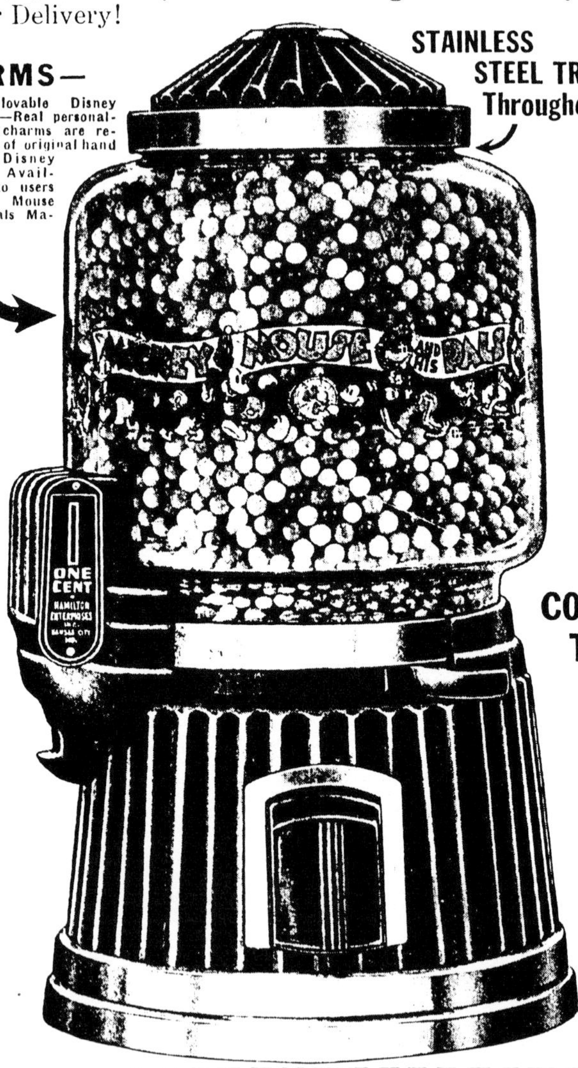

MAIL COUPON Today!

12 Outstanding Features:

1. Stainless Steel trim won't rust.
2. Best Balanced machine ever offered.
3. Black porcelain enamel base and top. Easy to clean —stays new looking indefinitely.
4. Beautiful in appearance—symmetrical and artistic.
5. Equipped with effective slug ejector that eliminates 95% of slug trouble.
6. Most satisfactory precision mechanism yet developed.
7. Mechanism guaranteed against defects in workmanship for 5 years.
8. Full six-pound bulk candy capacity.
9. Positive delivery—no failures.
10. Mechanism instantly adjustable to vend any kind of candy, peanuts or confection.
11. Suction pads on bottom hold machine securely to counter.
12. Design and every detail of machine personally inspected and approved by Walt Disney and associates.

BONA-FIDE DISTRIBUTORS WANTED... LIVE-WIRE SALESMEN WIRE OR WRITE!

USE THE COUPON NOW—Fill in, tear out and mail the coupon today—get all the facts about the fastest 1¢ vendor ever built—and sure-fire merchandising plan.

HAMILTON ENTERPRISES, Inc.
727 Holmes St., Kansas City, Mo.

Gentlemen: - I want to increase my earnings 100%. Please give me full details.

Name ______________________

Address ______________________

____

HAMILTON ENTERPRISES INC.

727 HOLMES ST. **A. T. HAMILTON, *President*** **KANSAS CITY, MO.**

AD-LEE ANNOUNCES A NEW MACHINE
. . . . AND A NEW POLICY

WE DEFINITELY WILL NOT FILL ANY INITIAL ORDERS FOR MORE THAN ONE MACHINE. WE ARE SOLD ON THE MERITS OF OUR PRODUCT. WE FURTHER KNOW THAT YOU WILL MAKE MORE MONEY OPERATING AD-LEE'S THREE COMPARTMENT (PENNY SHOP) MERCHANDISER THAN YOU HAVE EVER REALIZED ON ANY OTHER BULK VENDOR. WE DEMAND THAT YOU PROVE THESE FACTS TO YOUR ENTIRE SATISFACTION BEFORE YOU INVEST IN MORE THAN ONE MACHINE.

**ONE-THIRD DEPOSIT
BALANCE C. O. D.**

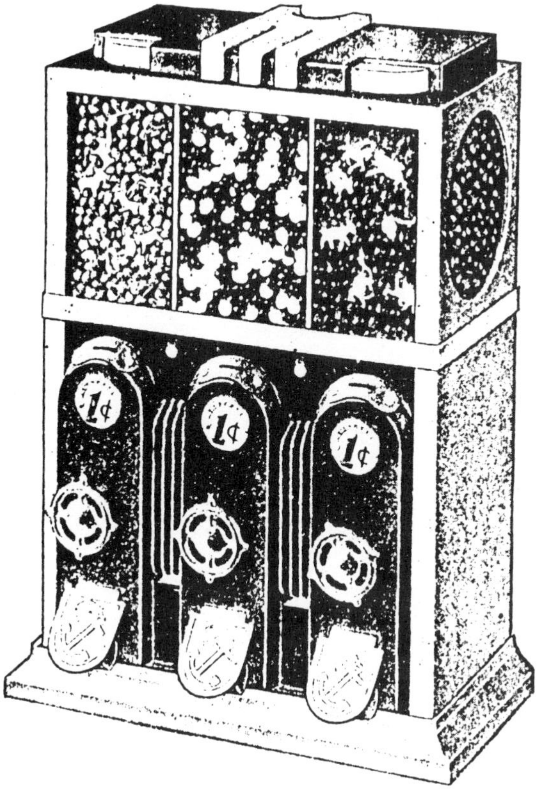

- FULL VISION DISPLAY
- TROUBLE FREE MECHANISM!
- MAXIMUM SLUG PROTECTION!
- STURDY CONSTRUCTION!
- RELIABLE GUARANTEE AGAINST DEFECTS IN MATERIAL AND WORKMANSHIP!
- COMPACT—18″ HIGH, 13″ WIDE, 8″ DEEP. CAPACITY APPROXIMATELY 5½ LBS. PER COMPARTMENT — TOTAL CAPACITY APPROXIMATELY 16½ LBS. WEIGHT (EMPTY) 39 LBS. REQUIRES ONLY 13″ OF COUNTER SPACE.
- FRONT AND BASE CAST IRON PORCELAIN ENAMEL, EVERLASTING.

ORDER ONE MACHINE TODAY. PRICE ONLY $29.50 CHICAGO F. O. B.

THE AD-LEE COMPANY - *Manufactuers*
821 SO. WABASH AVE. **CHICAGO, ILL.**

1939 SENSATIONS!

5 STAR FINAL

brings in the pennies; perfect for CLOSED LOCATIONS! No money ever in sight. Skill game that pays liberal odds in national confections. Sturdy pearline baked enamel; 5-tumbler lock.

SAMPLE **$3.95**
(Cheaper in Quantities)

½ Dep. Bal. C.O.D.

Watch pennies skip and skate for a BIGGER TAKE!

SELMOR MODEL "E"
Streamlined baked enamel and chromium; vends all candies, nuts, charms, ball gum for BIGGER PROFITS!
SAMPLE.......... **$6.25**
(Cheaper in Quantities)
Guaranteed 5 YEARS!
SLUG REJECTOR
SAMPLE $1.50 Extra
(Quantities cheaper)
Magnetic; stops chiselers. Fits all side slot machines: Silver King, Universal, SELMOR, Columbus, etc. Easy to install.

No Personal Checks, Please

GREAT STATES MFG. CO.
Dept. J, 1601-07 East 39th St.,
KANSAS CITY, MO.

The NEW ERA VENDORS

TWO COMPARTMENT
$11.00 each

Do not let the low price deceive you

Our claim is that these are the best compartment Vendors on the market regardless of price. Many new features never found before on any Vendors at any price.

Being the largest operator of compartment Vendors in Western Penna., we know the needs of an operator. Two years in the making, and by entire new method of manufacturing. Because of our low overhead we can offer you better Vendors at unheard low prices.

The Penny Mechanism is 90% slug proof, practically impossible for it to jam. One screw removes the machanism from the cabinet, and adjustments are made without taking the mechanism apart. Vends all type of bulk merchandise such as pistachios, candy with charms, etc.

THREE COMPARTMENT
$16.00 each

HERE IS OUR CHALLENGE

Order a sample machine, try it out for 30 to 60 days; if you are not satisfied, you may return the machine and receive your full purchase price back.

The PENNY KING Company

L. O. HARDMAN, *Sole Owner*

2418 Transport Street Pittsburgh, Pa.

Single Commander

Triple Commander

Double Commander

These beautiful—inviting vendors COMMAND the best locations. They vend nuts, pistachios, confections and charms. They have fully adjustable portion delivery. Made of NON-TARNISHABLE ALUMINUM that will stay bright like new. They are easy to keep clean and invite patronage.

Better locations may be secured when you can offer BETTER EQUIPMENT. A larger income may be earned because these machines GET THE PLAY. Only a screw driver is needed to service the COMMANDER LINE. (Because of the shortage of aluminum which is required for defense, we can promise delivery only on our present supply of a few hundred of these machines.)

Prices are low enough to justify replacing your old and obsolete machines with these up-to-date vendors. A liberal allowance for your old machines in trade.

Flash-A-Light

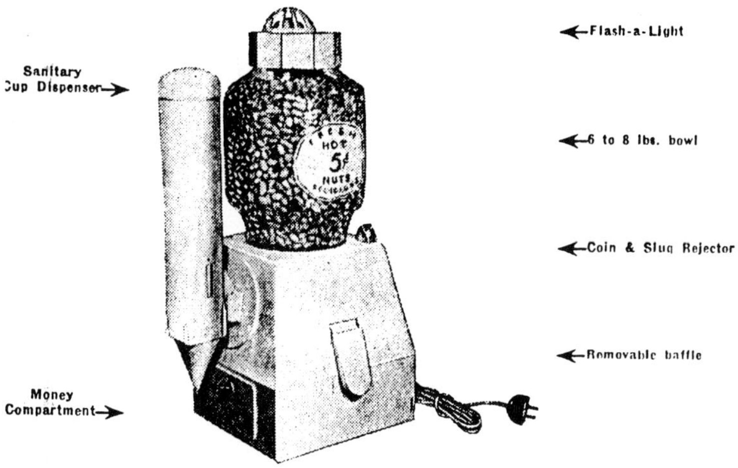

IT'S HERE — THE NEW SENSATIONAL 5c FLASH-A-LIGHT HOT NUT VENDOR

FLASH-A-LIGHT— the automatic machine that operators and merchants are clamoring for.

FLASH-A-LIGHT— the machine that dispenses all types of nuts such as: Cashews, Pistachios, Almonds, Filberts, Mixed Nuts, Virginia Peanuts, Indian Nuts. The supply of these nuts is guaranteed.

FLASH-A-LIGHT— the machine that is equipped with a National Slug Rejector which is a vital feature that prevents jamming of machine by foreign materials, and guarantees continuous operation of machine.

FLASH-A-LIGHT— the machine that is equipped with a glass bowl that dispenses 6 to 8 lbs. of various nuts—a pleasure for the operator or the merchant.

FLASH-A-LIGHT— the machine that is equipped with a removable baffle which can be cleaned in order to prevent rancidity and to adjust to the proper amount of nuts to be dispensed. Actual time to adjust baffle, clean, fill, and service machine on location is five minutes.

FLASH-A-LIGHT— the attractive gleaming white enamel sanitary hot nut machine which will add a decorative touch to any location. All machines are equipped with Sanitary Cup Dispensers and rubber suction cups on base of machine.

FLASH-A-LIGHT— the machine that is equipped with a guaranteed heating unit and an electric extension cord for AC-DC current.

FLASH-A-LIGHT— the machine that is equipped with a large money compartment and is protected by a National lock.

FLASH-A-LIGHT— the machine that is equipped with a colored bulb on top that flashes on and off to attract continuous attention—IT MUST BE SEEN!

AND REMEMBER THIS! — for real money making profit and tremendous earnings, it's *FLASH-A-LIGHT, the ideal operator's machine.*

Territories are now available for financially wide-awake experienced salesmen or distributors. Only those who can maintain a monthly quota basis will be given an exclusive distributorship in these territories.

DON'T WAIT! ACT NOW!

Wire, Write or Phone

AUTOMATIC MACHINE CO. OF NEW YORK
33 West 60 Street, New York 23, New York
Columbus 5-6464-5-6

WANTED

"LIVE WIRE"

DISTRIBUTORS

for

THE 2 IN 1 VENDOR

"America's Finest Merchandiser"

Must be financially able to operate as well as sell routes of our 2 In 1 Vendors. We are not seeking jobbers who merely carry a sample machine and do not promote sales. Distributors who follow our successful sales plan, earn good profits while operating and when selling. Exclusive territory allotted to producers. Write or wire today provided you meet our requirements.

"Sales Manager"

D. Robbins & Co., Mfrs.
1141-J DeKalb Ave.
BROOKLYN, NEW YORK

BUREL VENDERS

FIRST... Save You Money
THEN... Make You Money

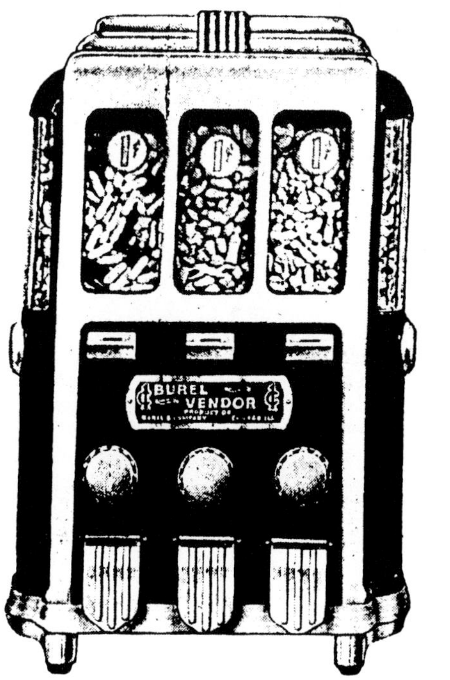

— LEADERSHIP —

because we PRACTICE what we PREACH and BUREL VENDERS make money for us on location. Why not duplicate our success?

Originators of the Summer Salted Nut Campaign.

Write for Details

BUREL & CO. 675-679 Orleans St. CHICAGO, ILL.

NU WAY VENDERS

for Peanuts and Small Confections

For use in Beer Gardens, Drug Stores, Restaurants, etc. Sturdy, well built and attractive machines. Entirely Hygienic as peanuts or confections drop into hand — not necessary to insert fingers to get them. Made with table or wall brackets, or counter stand. Money makers right from the minute they are installed.

We specialize in the manufacture of merchandise vending machines including American Cigarette Vendors.

WRITE FOR COMPLETE INFORMATION AND PRICES

Andrew Gorretta & Co.
6611 Euclid Avenue
CLEVELAND - - OHIO

INCREASING YOUR INCOME—

How to Operate the Watling Manufacturing Company's Springless Scales, with Vending Machines attached, in locations too far from your base of operations to service the Vending Machines.

No Springs
Style 14

NEW IMPROVEMENTS

All the newest improvements can be put on our oldest models of Springless Scales, therefore your scales never become obsolete.

Height over all	72 inches
Diameter of Head	24 inches
Width of Base	12 inches
Length of Base	26 inches

Net Weight—225 Pounds

1st—You can sell the merchandise to the storekeeper at a small profit, have him service the Vending Machines, and keep all the net profit on the Vending Machines.

2nd—You to keep the gross earnings of the Scale for your profit.

3rd—The earnings of the Vending Machines will be increased over FIFTY PER CENT when attached to the Scale.

4th—The earnings of the Scale will be increased over TWENTY-FIVE PER CENT with the Vending Machines attached.

5th—By giving the location all the profits on the Vending Machines, it induces the proprietor to keep the Scale and Machines clean and to place the Scale in a prominent location in his store, and absolutely eliminates competition.

6th—We furnish brackets for our Scales that enable you to use any style of Vending Machine you care to.

7th—We were the first to attach Vending Machines to our Scales and now, after proving the practicability of the combination for over 50 years, we strongly urge their use.

INVENTED AND MADE ONLY BY

WATLING MANUFACTURING CO.

701-707 W. Washington Blvd., Chicago, Illinois

HITTING THE VENDING MACHINE FIELD

LI'L ABNER
•
DAISY MAE
•
VENDAR-BAR

THE NEW
MERCHANDISE VENDOR

Specifications for
VENDAR-BAR . . .

Capacity . . . 104 Packages
Made of Heavy Gauge Steel
Baked Enamel Finish
Inside Parts Plated
Weight 35½ Pounds
8-Column Selective Type Merchandiser
Height, 23 Inches
Height With Stand, 55 In.
Width, 12½ Inches
Depth, 11½ Inches
National Coin Rejector

10¢ Coin Chute
Coin Scavenger
Wall Type
Counter Type
Stand Type (Optional)
In or Outside Locations
Sells 10¢ Pkg. of Delicious Chocolate Coated Raisins
Fresh Raisins
Fresh Spanish Peanuts
Fresh Assorted Caramels

50 MILLION DAILY READERS OF LI'L ABNER COMIC STRIP. THIS IS THE MOST SENSATIONAL MACHINE IN YEARS.

PUBLICITY AND PROMOTION

200 Radio and TV Personal Appearances Yearly
Daily Newspaper Mention — Editorial and Column
Two National Magazine Articles Monthly
Macy's, Bamberger's and Easter Parades
Ringling Bros. and Barnum & Bailey Circus
Lena the Hyena Contest — One Million Entries
Shmoo Birthday Celebrations
National Coverage on Use of George Jessel, Arthur Godfrey, Kate Smith, Frank Sinatra, Drew Pearson, etc., in strip

• FRANCHISES OPEN •

CALIFORNIA DISPENSER CO.

8134 W. 3RD STREET
LOS ANGELES 48, CALIFORNIA

TARGET PRACTICE MACHINE

Far in the Lead as a REAL MONEY MAKER. Shoot the Penny, Which Is Returned if Bull's-eye Is Scored.

Purely a game of skill. Does not violate the laws. Look at these figures of receipts from single machine:

RENFROE'S DRUG STORE, 29 Days $43.10
THE CHOCOLATE SHOP, 3 Days 7.21
RED CROSS PHARMACY, 63 Days 98.05

Rush your order in for one or one hundred at once. Single Machines **$20.00 Each.**

10 Machines at	$15.00	50 Machines at	$11.00
25 Machines at	12.50	100 or More at	10.00

Dimensions: Length, 16 inches; Height, 9½ inches; Width, 6 inches; Weight, 8½ lbs.

In order to get you to try one of these money-makers, will make you a special price on sample machine of **$12.50**, cash in full with order. ONLY THE ONE MACHINE TO EACH CUSTOMER AT THIS PRICE.

SPECIAL 30-DAY OFFER: TEN MACHINES AT THE 100-LOT PRICE.

Get exclusive operating rights for your territory. For Wisconsin territory write
Z. E. BUZZELL, 1681 24th Street, Milwaukee, Wisconsin.
You cannot afford to pass this up. Write quick.

THE SPECIALTY MANUFACTURING CO., 123 South Jefferson St., Chicago, Ill.

ADVANCE

Selective Vendor
Gum—Mints

Hot Peanut Machine

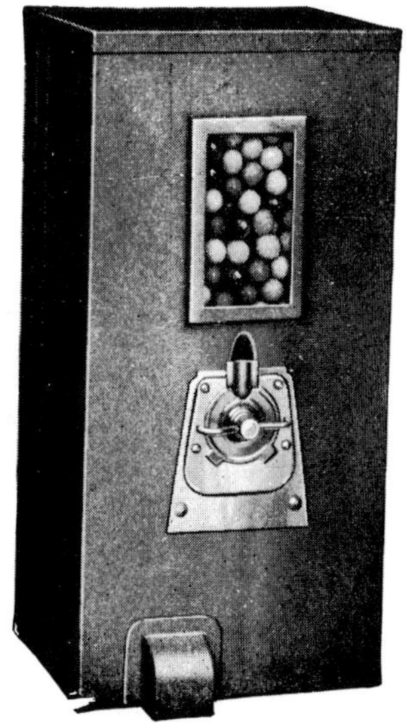

Operator's Special
Ball Gum—Peanuts

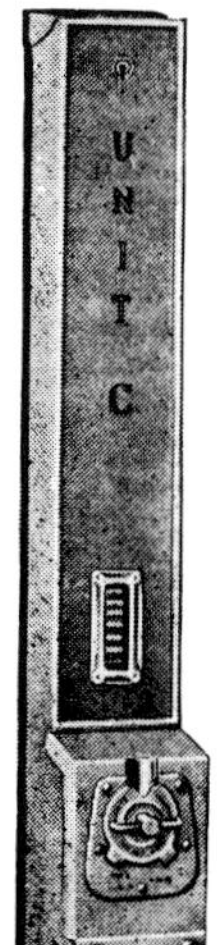

Unit-C
Gum - Mints, Etc.

Advance "Duo"

Candy Pellet Machine

A

B

C

I

J

K

L

M

N